# Anisotropic Hardy Spaces and Wavelets

# MEMOIRS
## of the
## American Mathematical Society

Number 781

# Anisotropic Hardy Spaces and Wavelets

Marcin Bownik

July 2003 • Volume 164 • Number 781 (third of 5 numbers) • ISSN 0065-9266

**American Mathematical Society**
Providence, Rhode Island

2000 *Mathematics Subject Classification.* Primary 42B30, 42C40; Secondary 42B20, 42B25.

---

**Library of Congress Cataloging-in-Publication Data**

Bownik, Marcin, 1971–
　　Anisotropic Hardy spaces and wavelets / Marcin Bownik.
　　　p. cm. — (Memoirs of the American Mathematical Society, ISSN 0065-9266 ; no. 781)
　　"Volume 164, number 781 (third of 5 numbers)."
　　Includes bibliographical references and index.
　　ISBN 0-8218-3326-X (alk. paper)
　　1. Hardy spaces. 2. Wavelets (Mathematics) I. Title. II. Series.

QA3.A57　no. 781
[QA331.5]
510 s–dc21
[515′.94]　　　　　　　　　　　　　　　　　　　　　　　　　　　　　　　　2003048023

---

## Memoirs of the American Mathematical Society

*This journal is devoted entirely to research in pure and applied mathematics.*

　　**Subscription information.** The 2003 subscription begins with volume 161 and consists of six mailings, each containing one or more numbers. Subscription prices for 2003 are $555 list, $444 institutional member. A late charge of 10% of the subscription price will be imposed on orders received from nonmembers after January 1 of the subscription year. Subscribers outside the United States and India must pay a postage surcharge of $31; subscribers in India must pay a postage surcharge of $43. Expedited delivery to destinations in North America $35; elsewhere $130. Each number may be ordered separately; *please specify number* when ordering an individual number. For prices and titles of recently released numbers, see the New Publications sections of the *Notices of the American Mathematical Society*.
　　**Back number information.** For back issues see the *AMS Catalog of Publications*.
　　Subscriptions and orders should be addressed to the American Mathematical Society, P. O. Box 845904, Boston, MA 02284-5904, USA. *All orders must be accompanied by payment.* Other correspondence should be addressed to 201 Charles Street, Providence, RI 02904-2294, USA.
　　**Copying and reprinting.** Individual readers of this publication, and nonprofit libraries acting for them, are permitted to make fair use of the material, such as to copy a chapter for use in teaching or research. Permission is granted to quote brief passages from this publication in reviews, provided the customary acknowledgment of the source is given.
　　Republication, systematic copying, or multiple reproduction of any material in this publication is permitted only under license from the American Mathematical Society. Requests for such permission should be addressed to the Acquisitions Department, American Mathematical Society, 201 Charles Street, Providence, Rhode Island 02904-2294, USA. Requests can also be made by e-mail to **reprint-permission@ams.org**.

---

　　*Memoirs of the American Mathematical Society* is published bimonthly (each volume consisting usually of more than one number) by the American Mathematical Society at 201 Charles Street, Providence, RI 02904-2294, USA. Periodicals postage paid at Providence, RI. Postmaster: Send address changes to Memoirs, American Mathematical Society, 201 Charles Street, Providence, RI 02904-2294, USA.

© 2003 by the American Mathematical Society. All rights reserved.
This publication is indexed in *Science Citation Index*®, *SciSearch*®, *Research Alert*®, *CompuMath Citation Index*®, *Current Contents*®/*Physical, Chemical & Earth Sciences*.
Printed in the United States of America.

∞ The paper used in this book is acid-free and falls within the guidelines established to ensure permanence and durability.
Visit the AMS home page at http://www.ams.org/

10 9 8 7 6 5 4 3 2 1　　08 07 06 05 04 03

# Contents

Chapter 1.  Anisotropic Hardy Spaces

1. Introduction  1
2. The space of homogeneous type associated with the discrete group of dilations  5
3. The grand maximal definition of anisotropic Hardy spaces  11
4. The atomic definition of anisotropic Hardy spaces  19
5. The Calderón-Zygmund decomposition for the grand maximal function  23
6. The atomic decomposition of $H^p$  35
7. Other maximal definitions  42
8. Duals of $H^p$  50
9. Calderón-Zygmund singular integrals on $H^p$  60
10. Classification of dilations  70

Chapter 2.  Wavelets

1. Introduction  82
2. Wavelets in the Schwartz class  85
3. Limitations on orthogonal wavelets  89
4. Non-orthogonal wavelets in the Schwartz class  94
5. Regular wavelets as an unconditional basis for $H^p$  96
6. Characterization of $H^p$ in terms of wavelet coefficients  104

Notation Index  116

Bibliography  118

# Abstract

In this paper, motivated in part by the role of discrete groups of dilations in wavelet theory, we introduce and investigate the anisotropic Hardy spaces associated with very general discrete groups of dilations. This formulation includes the classical isotropic Hardy space theory of Fefferman and Stein and parabolic Hardy space theory of Calderón and Torchinsky.

Given a *dilation* $A$, that is an $n \times n$ matrix all of whose eigenvalues $\lambda$ satisfy $|\lambda| > 1$, define the radial maximal function

$$M_\varphi^0 f(x) := \sup_{k \in \mathbb{Z}} |(f * \varphi_k)(x)|, \qquad \text{where } \varphi_k(x) = |\det A|^{-k} \varphi(A^{-k} x).$$

Here $\varphi$ is any test function in the Schwartz class with $\int \varphi \neq 0$. For $0 < p < \infty$ we introduce the corresponding anisotropic Hardy space $H_A^p$ as a space of tempered distributions $f$ such that $M_\varphi^0 f$ belongs to $L^p(\mathbb{R}^n)$.

Anisotropic Hardy spaces enjoy the basic properties of the classical Hardy spaces. For example, it turns out that this definition does not depend on the choice of the test function $\varphi$ as long as $\int \varphi \neq 0$. These spaces can be equivalently introduced in terms of grand, tangential, or nontangential maximal functions. We prove the Calderón-Zygmund decomposition which enables us to show the atomic decomposition of $H_A^p$. As a consequence of atomic decomposition we obtain the description of the dual to $H_A^p$ in terms of Campanato spaces. We provide a description of the natural class of operators acting on $H_A^p$, i.e., Calderón-Zygmund singular integral operators. We also give a full classification of dilations generating the same space $H_A^p$ in terms of spectral properties of $A$.

In the second part of this paper we show that for every dilation $A$ preserving some lattice and satisfying a particular expansiveness property there is a multiwavelet in the Schwartz class. We also show that for a large class of dilations (lacking this property) all multiwavelets must be combined minimally supported in frequency, and thus far from being regular. We show that $r$-regular (tight frame) multiwavelets form an unconditional basis (tight frame) for the anisotropic Hardy space $H_A^p$. We also describe the sequence space characterizing wavelet coefficients of elements of the anisotropic Hardy space.

---

Received by the editor May 15, 2000.

2000 *Mathematics Subject Classification*. Primary 42B30, 42C40; Secondary 42B20, 42B25.

*Key words and phrases*. Anisotropic Hardy space, radial maximal function, nontangential maximal function, grand maximal function, Calderón-Zygmund decomposition, atomic decomposition, Calderón-Zygmund operator, Campanato space, wavelets, frame, unconditional basis.

The author thanks his advisor, Prof. Richard Rochberg, for his support, guidance and encouragement.

CHAPTER 1

# Anisotropic Hardy spaces

## 1. Introduction

In the first chapter of this monograph we develop the real variable theory of Hardy spaces $H^p$ ($0 < p < \infty$) on $\mathbb{R}^n$ in what we believe is greater generality than has ever been done.

**Historical Background.** The theory of Hardy spaces is very rich with many highly developed branches. A recent inquiry in MathSciNet® (the database of *Mathematical Reviews* since 1940) revealed a fast growing collection of more than 2800 papers related to some extent to various Hardy spaces. Therefore, we can sketch only the most significant highlights of this theory.

Initially, Hardy spaces originated in the context of complex function theory and Fourier analysis in the beginning of twentieth century. The classical Hardy space $H^p$, where $0 < p < \infty$, consists of holomorphic functions $f$ defined on the unit disc such that

$$||f||_{H^p} := \sup_{0<r<1} \left[ \int_0^1 |f(re^{2\pi i\theta})|^p d\theta \right]^{1/p} < \infty,$$

or on the upper half plane such that

$$||f||_{H^p} := \sup_{0<y<\infty} \left[ \int_{-\infty}^\infty |f(x+iy)|^p dx \right]^{1/p} < \infty.$$

If $p = \infty$ we replace the above integrals by the suprema. For a systematic exposition of the subject see books by Duren [Du], Garnett [Ga], and Koosis [Ko].

The possible generalizations of these spaces to higher dimensions include Hardy spaces on the unit ball in $\mathbb{C}^n$, on the polydisc or on the tube domains over cones, see books of Rudin [Ru1, Ru2], and Stein and Weiss [SW2]. Another possibility is to consider spaces of conjugate harmonic functions $f = (u_0, \ldots, u_n)$ on $\mathbb{R}^n \times (0, \infty)$, satisfying certain natural generalizations of the Cauchy-Riemann equations and the size condition

$$||f||_{H^p} := \sup_{0<y<\infty} \left[ \int_{\mathbb{R}^n} |f(x_1, \ldots, x_n, y)|^p dx_1 \ldots dx_n \right]^{1/p} < \infty,$$

see Stein and Weiss [SW1, SW2]. In this development the attention is focused on the boundary values of the harmonic functions, which are distributions on $\mathbb{R}^n$. The harmonic functions can be then recovered from the boundary values by Poisson integral formula. The resulting spaces $H^p(\mathbb{R}^n)$ are equivalent to $L^p(\mathbb{R}^n)$ for $p > 1$. However, for $p \leq 1$ these spaces differ from $L^p(\mathbb{R}^n)$ and are better suited for the purposes of harmonic analysis than $L^p(\mathbb{R}^n)$. Indeed, singular integral operators and multiplier operators turn out to be bounded on $H^p$, see Stein [St1].

The beginning of the 1970's marked the birth of real-variable theory of Hardy spaces as we know it today. First, Burkholder, Gundy, and Silverstein in [BGS] using Brownian motion methods showed that $f$ belongs to the classical Hardy space $H^p$ if and only if the nontangential maximal function of Re $f$ belongs to $L^p$. The real breakthrough came in the work of C. Fefferman and Stein [FS2]. They showed that $H^p$ in $n$ dimensions can be defined as the space of tempered distributions $f$ on $\mathbb{R}^n$ whose radial maximal function $M_\varphi^0$ or nontangential maximal function $M_\varphi$ belong to $L^p(\mathbb{R}^n)$, where

$$M_\varphi^0 f(x) := \sup_{0<t<0} |(f * \varphi_t)(x)|,$$
$$M_\varphi f(x) := \sup_{0<t<\infty} \sup_{|x-y|<t} |(f * \varphi_t)(y)|,$$

and $\varphi_t(x) := t^{-n}\varphi(x/t)$. Here $\varphi$ is any test function in the Schwartz class with $\int \varphi \neq 0$ or the Poisson kernel $\varphi(x) = (1+|x|^2)^{-(n+1)/2}$ (in this case $f$ is restricted to bounded distribution) and the definition of $H^p$ does not depend on this choice. To prove this impressive result C. Fefferman and Stein introduced a very important tool, the grand maximal function, which can also be used to define Hardy spaces $H^p$. The real analysis methods also played a decisive role in the well-known C. Fefferman duality theorem between $H^1$ and $BMO$—the space of functions of bounded mean oscillation.

Further insight into the theory of Hardy spaces came from the works of Coifman [Co] ($n = 1$) and Latter [La] ($n \geq 1$) where the atomic decomposition of elements in $H^p(\mathbb{R}^n)$ ($p \leq 1$) was exhibited. Atoms are compactly supported functions satisfying certain boundedness properties and some number of vanishing moments. The Hardy space $H^p(\mathbb{R}^n)$ can be thought of in terms of atoms and many important theorems can be reduced to easy verifications of statements for atoms.

Other developments followed. Coifman and Weiss in [CW2] introduced Hardy spaces $H^p$ for the general class of spaces of homogeneous type using as a definition atomic decompositions. Since there is no natural substitute for polynomials, the Hardy spaces $H^p$ on spaces of homogeneous type can be defined only for $p \leq 1$ sufficiently close to 1. Another approach started in the work of Calderón and Torchinsky [CT1, CT2] who developed theory of Hardy spaces $H^p$ ($0 < p < \infty$) on $\mathbb{R}^n$ for nonisotropic dilations. The theory of Hardy spaces was also established on more general groups than $\mathbb{R}^n$. For the Heisenberg group it was done by Geller [Ge] and for general homogeneous groups by Folland and Stein [FoS]. We also mention the development of Hardy space on subsets of $\mathbb{R}^n$ by Jonsson and Wallin [JW] and weighted Hardy spaces on $\mathbb{R}^n$ by Strömberg and Torchinsky in [ST1, ST2].

**Parabolic Hardy spaces.** Calderón and Torchinsky initiated the study of Hardy spaces on $\mathbb{R}^n$ with nonisotropic dilations in [CT1, CT2]. They start with a one parameter continuous subgroup of $GL(\mathbb{R}^n, n)$ of the form $\{A_t : 0 < t < \infty\}$ satisfying $A_t A_s = A_{ts}$ and

$$t^\alpha |x| \leq |A_t x| \leq t^\beta |x| \qquad \text{for all } x \in \mathbb{R}^n, t \geq 1,$$

for some $1 \leq \alpha \leq \beta < \infty$. The infinitesimal generator $P$ of $A_t = t^P := \exp(P \ln t)$ satisfies $\langle Px, x \rangle \geq \langle x, x \rangle$, where $\langle \cdot, \cdot \rangle$ is the standard scalar product in $\mathbb{R}^n$. The induced nonisotropic norm $\rho$ on $\mathbb{R}^n$ satisfies $\rho(A_t x) = t\rho(x)$. The parabolic Hardy space $H^p$ ($0 < p < \infty$) is defined as a space of tempered distributions $f$ whose nontangential function $M_\varphi f$ belongs to $L^p(\mathbb{R}^n)$, where

$$M_\varphi f(x) := \sup_{\rho(x-y)<t} |(f * \varphi_t)(x)|, \qquad \varphi_t(x) = t^{-\operatorname{tr} P} \varphi(A_t^{-1} x),$$

$\varphi$ is any test function with $\int \varphi \neq 0$. Calderón and Torchinsky also obtain equivalent formulations of parabolic Hardy spaces using the grand maximal functions, Luzin functions, and Littlewood-Paley functions. The atomic decomposition for theses spaces was done by Calderón [Ca, Ga] and Latter and Uchiyama [LU].

It is worth noting that the general setup for defining Hardy spaces on homogeneous groups developed by Folland and Stein [FoS] presupposes that the dilation group $\{A_t : 0 < t < \infty\}$ is of the form $A_t = \exp(P \ln t)$, where $P$ is a diagonalizable element of $GL(\mathbb{R}^n, n)$ with positive eigenvalues. In general, such a matrix $P$ need not satisfy $\langle Px, x \rangle \geq \langle x, x \rangle$. Conversely, generators $P$ allowed in [CT1, CT2] need not even be diagonalizable. Therefore, on the formal level, dilations structures considered in [CT1, CT2, FoS] do not in general overlap. As a consequence some Hardy spaces procured in one theory do not appear in the other, and vice versa.

The optimal solution would be to relax even further the assumptions on the group of dilations $\{A_t : 0 < t < \infty\}$ by merely assuming that $\lim_{t \to 0^+} ||A_t|| = 0$. This is the approach we adapt in our work with the exception that we allow even more general *discrete* dilation structures which have originated in the theory of wavelets.

**Description of the chapter.** The scope of this chapter is an introduction and investigation of the real variable theory of Hardy spaces associated with a general group of dilations. By a group of dilations we mean a one parameter, discrete subgroup of $GL(\mathbb{R}, n)$, i.e., $\{A^k : k \in \mathbb{Z}\}$, where $A$ is a generating $n \times n$ matrix whose all eigenvalues $\lambda$ satisfy $|\lambda| > 1$. We investigate the properties of the space of homogeneous type induced by this group of dilations in Section 2. In the next section we define the anisotropic Hardy space $H_A^p(\mathbb{R}^n)$ as a space of tempered distributions $f$ whose grand maximal function belongs to $L^p(\mathbb{R}^n)$. The most straightforward definition of these spaces is obtained using the radial maximal function. That is, for a dilation $A$ and $0 < p < \infty$ we introduce the corresponding anisotropic Hardy space $H_A^p$ as a space of tempered distributions $f$ whose radial maximal function $M_\varphi^0 f$ given by

$$M_\varphi^0 f(x) = \sup_{k \in \mathbb{Z}} |(f * \varphi_k)(x)|, \qquad \text{where } \varphi_k(x) = |\det A|^{-k} \varphi(A^{-k} x),$$

belongs to $L^p$. Here $\varphi$ is any test function in the Schwartz class with $\int \varphi \neq 0$. By virtue of the main theorem in Section 7 this definition does not depend on the choice of $\varphi$ as long as $\int \varphi \neq 0$. In Section 4 we introduce anisotropic Hardy spaces by means of atomic decompositions. In Section 5, one of the longest and most technical sections, we carefully derive the Calderón-Zygmund decomposition. In Sections 6 and 7 we prove that various definitions of Hardy spaces in terms of grand, radial, tangential, and nontangential maximal functions and in terms of atomic decompositions are all equivalent. In the next section we describe the duals of anisotropic Hardy spaces. In Section 9 we develop the theory of Calderón-Zygmund operators acting on anisotropic $H_A^p$ spaces. Finally, in the last section we classify dilations which yield equivalent anisotropic Hardy spaces.

We want to emphasize that the presentation of Sections 3–7 is greatly influenced by the excellent exposition of Hardy spaces on homogeneous groups in the book of Folland and Stein [FoS].

## 2. The space of homogeneous type associated with the discrete group of dilations

The concept of a dilation is a fundamental one in our work.

DEFINITION 2.1. A *dilation* is $n \times n$ real matrix $A$, such that all eigenvalues $\lambda$ of $A$ satisfy $|\lambda| > 1$.

It is clear that $A$ is a dilation if and only if $||A^{-j}|| \to 0$ as $j \to \infty$. We could alternatively define a dilation as a matrix whose all eigenvalues $\lambda$ satisfy $0 \neq |\lambda| < 1$. The inverse of this matrix becomes then a dilation in the sense of Definition 2.1.

For any dilation $A$ we consider the corresponding discrete group of linear transformations $\{A^j : j \in \mathbb{Z}\}$ which induces a natural structure of a space of homogeneous type on $\mathbb{R}^n$. For general facts about spaces of homogeneous type we refer the reader to [Ch, CW2, HS, MS1–MS3]. Unless the dilation $A$ is very special (to be made precise later) this structure is different than the usual isotropic structure of $\mathbb{R}^n$. In this section we present the underlying ideas.

To start we suppose $\lambda_1, \ldots, \lambda_n$ are eigenvalues of $A$ (taken according to the multiplicity) so that $1 < |\lambda_1| \leq \ldots \leq |\lambda_n|$. Let $\lambda_-$, $\lambda_+$ be any numbers so that $1 < \lambda_- < |\lambda_1| \leq |\lambda_n| < \lambda_+$. Then we can find a constant $c > 0$ so that for all $x \in \mathbb{R}^n$ we have

(2.1) $$1/c\lambda_-^j |x| \leq |A^j x| \leq c\lambda_+^j |x| \qquad \text{for } j \geq 0,$$

(2.2) $$1/c\lambda_+^j |x| \leq |A^j x| \leq c\lambda_-^j |x| \qquad \text{for } j \leq 0,$$

where $|\cdot|$ is a standard norm in $\mathbb{R}^n$. Note that (2.2) is a consequence of (2.1) and vice versa. Furthermore, if for any eigenvalue $\lambda$ of $A$ with $|\lambda| = |\lambda_1|$ (or $|\lambda| = |\lambda_n|$) the matrix $A$ does not have Jordan blocks corresponding to $\lambda$, i.e., $\ker(A - \lambda I) = \ker(A - \lambda I)^2$ then we can set $\lambda_- = |\lambda_1|$ ($\lambda_+ = |\lambda_n|$).

A set $\Delta \subset \mathbb{R}^n$ is said to be an *ellipsoid* if

(2.3) $$\Delta = \{x \in \mathbb{R}^n : |Px| < 1\},$$

for some nondegenerate $n \times n$ matrix $P$, where $|\cdot|$ denotes the standard norm in $\mathbb{R}^n$.

In general, we can not expect that the dilation $A$ is expansive in the standard norm, i.e., $|Ax| \geq |x|$ for all $x \in \mathbb{R}^n$. Nevertheless, by [Sz, Lemma 1.5.1], there exists a scalar product with the induced norm $|\cdot|_*$ and $r > 1$ so that

(2.4) $$|Ax|_* \geq r|x|_* \qquad \text{for } x \in \mathbb{R}^n.$$

We present the proof of this result for the sake of completeness.

LEMMA 2.2. *Suppose $A$ is a dilation. Then there exists an ellipsoid $\Delta$ and $r > 1$ such that*

(2.5) $$\Delta \subset r\Delta \subset A\Delta.$$

PROOF. Define the inner product $\langle \cdot, \cdot \rangle_*$ by

$$\langle x, y \rangle_* = \langle x, y \rangle + \langle A^{-1}x, A^{-1}y \rangle + \ldots + \langle A^{-k}x, A^{-k}y \rangle,$$

where $k$ is an integer satisfying $k > 2\ln c/\ln\lambda_-$, and $c$, $\lambda_-$ are as in (2.1). We claim that the norm $|x|_* = \langle x,x\rangle_*^{1/2}$ satisfies (2.4). Indeed, by (2.1) and (2.2)

$$|Ax|_*^2 = |Ax|^2 + |x|^2 + \ldots + |A^{-k+1}x|^2 = |x|_*^2 + |Ax|^2 - |A^{-k}x|^2$$
$$\geq |x|_*^2 + 1/c^2|x|^2 - c^2\lambda_-^{-2k}|x|^2 = |x|_*^2(1 + (c^{-2} - c^2\lambda_-^{-2k})|x|^2/|x|_*^2)$$
$$= |x|_*^2\left[1 + \frac{c^{-2} - c^2\lambda_-^{-2k}}{|x|^2 + \ldots + |A^{-k}x|^2}|x|^2\right] \geq |x|_*^2\left[1 + \frac{c^{-4} - \lambda_-^{-2k}}{1 + \ldots + \lambda_-^{-2k}}\right] = r^2|x|_*^2,$$

where $r$ is the square root of the last bracket. By a simple application of the Riesz Lemma there is a matrix $Q$ so that $\langle Qx, y\rangle = \langle x, y\rangle_*$. Clearly, $Q$ is self-adjoint and positive definite. If we take $P = Q^{1/2}$ then $|Px|^2 = \langle Qx, x\rangle = |x|_*$. Define $\Delta$ by (2.3), i.e., $\Delta = \{x \in \mathbb{R}^n : |x|_* < 1\}$. Since $|A^{-1}x|_* \leq |x|_*/r$ then $A^{-1}\Delta \subset r^{-1}\Delta$ and hence (2.5) holds. □

By a scaling we can additionally assume that ellipsoid $\Delta$ in Lemma 2.2 satisfies $|\Delta| = 1$. We define a family of balls around the origin as the sets

(2.6) $$B_k := A^k\Delta \quad \text{for } k \in \mathbb{Z}.$$

By (2.5) we have

(2.7) $\quad B_k \subset rB_k \subset B_{k+1}, \quad |B_k| = b^k, \quad$ where $b := |\det A| > 1$.

Even though the choice of the expansive ellipsoid $\Delta$ is not unique, from this point we fix one choice of $\Delta$, and consequently the $B_k$'s, for a given dilation $A$.

Next we introduce the natural concept of a quasi-norm which generalizes the usual norm on $\mathbb{R}^n$. A quasi-norm satisfies a discrete homogeneity property with respect to $A$ and a triangle inequality up to a constant.

DEFINITION 2.3. A *homogeneous quasi-norm* associated with a dilation $A$ is a measurable mapping $\rho : \mathbb{R}^n \to [0, \infty)$, so that
(i) $\rho(x) = 0 \iff x = 0$,
(ii) $\rho(Ax) = b\rho(x)$ for all $x \in \mathbb{R}^n$,
(iii) there is $c > 0$ so that $\rho(x + y) \leq c(\rho(x) + \rho(y))$ for all $x, y \in \mathbb{R}^n$.

It turns out that all quasi-norms associated to a fixed dilation $A$ are equivalent.

LEMMA 2.4. *Any two homogeneous quasi-norms $\rho_1$, $\rho_2$ associated with a dilation $A$ are equivalent, i.e., there exists a constant $c > 0$ so that*

(2.8) $$1/c\rho_1(x) \leq \rho_2(x) \leq c\rho_1(x) \quad \text{for } x \in \mathbb{R}^n.$$

PROOF. It suffices to show that for every quasi-norm $\rho$ we have

(2.9) $$0 < \inf_{x \in B_1\setminus B_0} \rho(x) \leq \sup_{x \in B_1\setminus B_0} \rho(x) < \infty.$$

Suppose on the contrary that $\sup_{x \in B_1\setminus B_0} \rho(x) = \infty$. Then we can find a sequence $(x_i) \subset \mathbb{R}^n$ such that $|x_i| \to 0$, $\rho(x_i) \to \infty$ as $i \to \infty$. Choose $M > 0$ so that the set

$$\Omega = \{x \in B_1 \setminus B_0 : \rho(x) \leq M\},$$

has measure strictly bigger than $(b-1)/2$. Choose $N$, so that $\rho(x_i) > 2cM$ for $i > N$. Since

$$\rho(x) \geq 1/c\rho(x_i) - \rho(x_i - x),$$

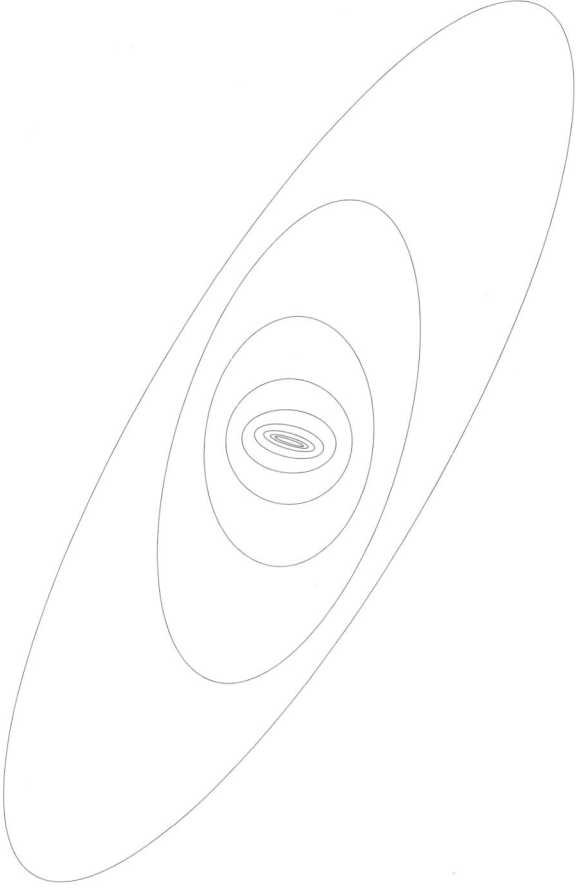

FIGURE 1. THE FAMILY OF DILATED BALLS $\{B_k : k \in \mathbb{Z}\}$.

we have $\rho(x) > M$ for $x \in x_i - \Omega$, i.e.,
$$(x_i - \Omega) \cap \Omega = \emptyset.$$
Since $-\Omega \subset B_1 \setminus B_0$ and $x_i \to 0$ we have
$$|(x_i - \Omega) \cap (B_1 \setminus B_0)| \to |\Omega| > (b-1)/2 \qquad \text{as } i \to \infty.$$
Hence,
$$b - 1 < |(x_i - \Omega) \cap (B_1 \setminus B_0)| + |\Omega| \leq |B_1 \setminus B_0| = b - 1,$$
for sufficiently large $i$, which is a contradiction.

Finally, suppose on the contrary that $\inf_{x \in B_1 \setminus B_0} \rho(x) = 0$. Then we can find a sequence $(x_i) \subset B_1 \setminus B_0$ such that $\rho(x_i) \to 0$ as $i \to \infty$. By selecting a subsequence we can assume $(x_i)$ converges to some point $x \neq 0$. Since
$$\rho(x) \leq c(\rho(x - x_i) + \rho(x_i)) \to 0 \qquad \text{as } i \to \infty,$$
we have $\rho(x) = 0$ which is a contradiction. Therefore (2.9) holds. □

The natural question is whether for two distinct dilations $A_1$ and $A_2$, the corresponding quasi-norms $\rho_1$ and $\rho_2$ are equivalent, i.e., (2.8) holds for some $c > 0$. One may wish to state the classification theorem in terms of eigenvalues and eigenspaces corresponding to Jordan blocks, which we postpone until Section 10, see Theorem 10.3. Instead, we consider the simplest situation.

EXAMPLE. Suppose $A = d\,\mathrm{Id}$ for some real $|d| > 1$. It is clear that $\rho(x) = |x|^n$ satisfies properties of the quasi-norm. Therefore all matrices of this form induce equivalent quasi-norms. Here it becomes clear why we impose $\rho(Ax) = b\rho(x)$ with $b = |\det A|$. It is not hard to see that any matrix $A$ without Jordan blocks and whose all of eigenvalues $\lambda$ satisfy $|\lambda| = d$ for some fixed $d > 1$ has the corresponding quasi-norm $\rho$ which is equivalent to $|\cdot|^n$. And vice versa, any other matrix whose all of eigenvalues are not equal in absolute value or with Jordan block(s) has a quasi-norm which is not equivalent to $|\cdot|^n$. Equivalently, a quasi-norm induced by a dilation $A$ is equivalent to $|\cdot|^n$ if and only if $A$ is diagonalizable over $\mathbb{C}$ with all eigenvalues equal in the absolute value. This provides the classification of dilations inducing the usual isotropic homogeneous structure on $\mathbb{R}^n$ announced in the beginning of this section.

For a fixed dilation $A$ we define the "canonical" quasi-norm used throughout this chapter.

DEFINITION 2.5. Define the *step homogeneous quasi-norm* $\rho$ on $\mathbb{R}^n$ induced by the dilation $A$ as

$$(2.10) \qquad \rho(x) = \begin{cases} b^j & \text{if } x \in B_{j+1} \setminus B_j \\ 0 & \text{if } x = 0. \end{cases}$$

Here $B_k = A^k \Delta$, where $\Delta$ is an ellipsoid from Lemma 2.2 and $|\Delta| = 1$.

Indeed, $\rho$ clearly satisfies (i) and (ii) of Definition 2.2. Let $\omega$ be the smallest integer so that $2B_0 \subset A^\omega B_0 = B_\omega$. The existence of $\omega$ is guaranteed by (2.5). Suppose $x, y \in \mathbb{R}^n$, $\rho(x) = b^i$, $\rho(y) = b^j$ for some $i, j \in \mathbb{Z}$. Thus, $x + y \in B_i + B_j \subset 2B_{\max(i,j)} \subset A^\omega B_{\max(i,j)}$. Therefore

$$\rho(x+y) \leq b^\omega b^{\max(i,j)} \leq b^\omega (b^i + b^j) = b^\omega (\rho(x) + \rho(y)),$$

and (iii) holds with the constant $c = b^\omega$.

Therefore, we can summarize that for each $i \in \mathbb{Z}$ we have

$$(2.11) \qquad \begin{aligned} x \in B_i \quad \text{and} \quad y \in B_i &\implies x + y \in B_{i+\omega}, \\ x \notin B_{i+\omega} \quad \text{and} \quad y \in B_i &\implies x + y \notin B_i. \end{aligned}$$

We also record two useful inequalities

$$(2.12) \qquad \max(1, \rho(x+y)) \leq b^\omega \max(1, \rho(x)) \max(1, \rho(y)) \qquad \text{for all } x, y, \in \mathbb{R}^n,$$
$$(2.13) \qquad \max(1, \rho(A^j x)) \leq b^j \max(1, \rho(x)) \qquad \text{for } j \geq 0, x \in \mathbb{R}^n.$$

REMARK. The notion of a quasi-norm put forth in Definition 2.3 seems to be missing in the literature. Lemarié-Rieusset in [LR] considered a closely related quasi-norm satisfying (i) and (ii) of Definition 2.3 which is $C^\infty$ on $\mathbb{R}^n$ except at the origin. He also gives a construction of such a quasi-norm for an arbitrary dilation $A$, see [LR]. Quasi-norms of Lemarié-Rieusset automatically satisfy the triangle inequality up to a constant. In our definition, which was motivated by

[LR], we give up the smoothness of quasi-norms to include the important class of step homogeneous quasi-norms used throughout this chapter.

We need two covering lemmas which hold in arbitrary spaces of homogeneous type, see [CW1, CW2]. However, we present them in a form adapted to our setting.

LEMMA 2.6 (WIENER). *Suppose $\Omega \subset \mathbb{R}^n$, and $r : \Omega \to \mathbb{Z}$ is an arbitrary function. Assume that either (a) $\Omega$ is bounded, or (b) $\Omega$ is open, $|\Omega| < \infty$, and $x + B_{r(x)} \subset \Omega$ for all $x \in \Omega$. Then there exists a sequence $(x_j) \subset \Omega$ (finite or infinite) so that the balls $x_j + B_{r(x_j)}$ are mutually disjoint and $\Omega \subset \bigcup_j (x_j + B_{r(x_j)+\omega})$.*

PROOF. In case (a), if $\sup_{x \in \Omega} r(x) = \infty$ then we can find $x \in \Omega$ so that $\Omega \subset x + B_{r(x)}$, and we are done. Thus, we can assume $\sup_{x \in \Omega} r(x) < \infty$. In case (b), since $|\Omega| < \infty$ we also must have $\sup_{x \in \Omega} r(x) < \infty$. Pick $x_1 \in \Omega$ such that $r(x_1) = \sup_{x \in \Omega} r(x)$. If $\Omega \subset x_1 + B_{r(x_1)+\omega}$ then we are done. Otherwise, we proceed inductively. Assume we have picked $x_1, \ldots, x_j$, set $\Omega' = \Omega \setminus \bigcup_{i=1}^{j}(x_i + B_{r(x_i)+\omega})$. If $\Omega = \emptyset$ we are done. If not, pick $x_{j+1} \in \Omega'$ such that $r(x_{j+1}) = \sup_{x \in \Omega'} r(x)$. Suppose $i < j$ and $(x_i + B_{r(x_i)}) \cap (x_j + B_{r(x_j)}) \neq \emptyset$ then $x_j - x_i \in B_{r(x_i)} - B_{r(x_j)} \subset 2B_{r(x_i)} \subset B_{r(x_i)+\omega}$, since $r(x_i) \geq r(x_j)$. Thus, $x_j \in B_{r(x_i)+\omega}$, which is a contradiction. Therefore, balls $x_j + B_{r(x_j)}$ are mutually disjoint.

If the sequence $(x_j)$ is finite then clearly $\Omega \subset \bigcup_j (x_j + B_{r(x_j)+\omega})$. If not we have $r(x_j) \to -\infty$ as $j \to \infty$ since $\Omega$ has finite measure and balls $x_j + B_{r(x_j)}$ are mutually disjoint. Suppose on the contrary that there exists $x \in \Omega$ such that $x \notin \bigcup_j (x_j + B_{r(x_j)+\omega})$. For sufficiently large $j$ we have $r(x) \geq r(x_j)$, which is a contradiction of the choice of $x_j$'s. □

LEMMA 2.7 (WHITNEY). *Suppose $\Omega \subset \mathbb{R}^n$ is open, and $|\Omega| < \infty$. For every integer $d \geq 0$ there exists a sequence of points $(x_j)_{j \in \mathbb{N}} \subset \Omega$, and a sequence of integers $(l_j)_{j \in \mathbb{N}} \subset \mathbb{Z}$, so that*
*(i) $\Omega = \bigcup (x_j + B_{l_j})$,*
*(ii) $x_j + B_{l_j - \omega}$ are pairwise disjoint for $j \in \mathbb{N}$,*
*(iii) for every $j \in \mathbb{N}$, $(x_j + B_{l_j + d}) \cap \Omega^c = \emptyset$, but $(x_j + B_{l_j + d + 1}) \cap \Omega^c \neq \emptyset$,*
*(iv) if $(x_i + B_{l_i + d - 2\omega}) \cap (x_j + B_{l_j + d - 2\omega}) \neq \emptyset$ then $|l_i - l_j| \leq \omega$,*
*(v) for each $j \in \mathbb{N}$, the cardinality of $\{i \in \mathbb{N} : (x_i + B_{l_i + d - 2\omega}) \cap (x_j + B_{l_j + d - 2\omega}) \neq \emptyset\}$ is less than $L$, where $L$ is a constant depending only on $d$.*

PROOF. For every $x \in \Omega$ define function
$$r(x) := \sup\{r \in \mathbb{Z} : x + B_{r+d+\omega} \subset \Omega\}.$$

Since $\Omega$ is open and has finite measure the supremum is always finite. By the Wiener Lemma applied to the function $r(x)$ we can find a sequence of $(x_j) \subset \Omega$ so that $(x_j + B_{r(x_j)})$ are mutually disjoint, and $(x_j + B_{r(x_j)+\omega}) \subset \Omega$ cover the set $\Omega$. Therefore, if we define $l_j := r(x_j) - \omega$ then (i) and (ii) hold. Clearly (iii) also holds by the choice of $r(x)$. If $d \geq 1$ then we have

$$\Omega = \bigcup_j (x_j + B_{l_j}) = \bigcup_j (x_j + B_{l_j + 1}) \supset \bigcup_j (x_j + rB_{l_j}),$$

where $r$ is the same as in Lemma 2.2, and therefore the sequence $(x_j)$ is necessarily infinite.

To show (iv), suppose $y \in (x_i + B_{l_i+d-2\omega}) \cap (x_j + B_{l_j+d-2\omega})$ and $l_j \geq l_i + \omega + 1$. Then
$$x_i - x_j \in y - B_{l_i+d-2\omega} - y + B_{l_j+d-2\omega} \subset B_{l_j+d-\omega}.$$
Since
$$x_i + B_{l_i+d+1} = (x_i - x_j) + x_j + B_{l_i+d+1} \subset x_j + B_{l_j+d-\omega} + B_{l_i+d+1} \subset x_j + B_{l_j+d},$$
we have by (iii)
$$\emptyset \neq (x_i + B_{l_i+d+1}) \cap \Omega^c \subset (x_j + B_{l_j+d}) \cap \Omega^c,$$
which is a contradiction of (iii). Therefore $l_j \leq l_i + \omega$, and by symmetry we obtain (iv).

Fix $j \in \mathbb{N}$ and consider
$$I = \{i \in \mathbb{N} : (x_i + B_{l_i+d-2\omega}) \cap (x_j + B_{l_j+d-2\omega}) \neq \emptyset\}.$$
If $i \in I$ then by (iv)
$$x_i + B_{l_i-\omega} \subset (x_i - x_j) + x_j + B_{l_i-\omega} \subset x_j + B_{l_i+d-2\omega} + B_{l_j+d-2\omega} + B_{l_i-\omega}$$
$$\subset x_j + B_{l_j+d-\omega} + B_{l_j+d-2\omega} + B_{l_j} \subset x_j + B_{l_j+d+\omega}.$$
Since for $i \in I$, $x_i + B_{l_j-2\omega} \subset x_i + B_{l_i-\omega}$ we conclude by (ii) that the cardinality of $I$ is less than
$$|B_{l_j+d+\omega}|/|B_{l_j-2\omega}| = b^{d+3\omega} = L. \qquad \square$$

## 3. The grand maximal definition of anisotropic Hardy spaces

DEFINITION 3.1. We say that a $C^\infty$ complex valued function $\varphi$ on $\mathbb{R}^n$ belongs to the Schwartz class $\mathcal{S}$, if for every multi-index $\alpha$ and integer $m \geq 0$ we have

$$(3.1) \qquad ||\varphi||_{\alpha,m} := \sup_{x \in \mathbb{R}^n} \rho(x)^m |\partial^\alpha \varphi(x)| < \infty.$$

The space $\mathcal{S}$ endowed with pseudonorms $||\cdot||_{\alpha,m}$ becomes a (locally convex) topological vector space. The dual space of bounded functionals on $\mathcal{S}$, i.e., the space of *tempered distributions* on $\mathbb{R}^n$, is denoted by $\mathcal{S}'$.

If $f \in \mathcal{S}'$ and $\varphi \in \mathcal{S}$ we denote the evaluation of $f$ on $\varphi$ by $\langle f, \varphi \rangle$. Sometimes we will pretend that distributions are functions by writing $\langle f, \varphi \rangle = \int f(x) \varphi(x) dx$. Convergence in $\mathcal{S}'$ will always mean weak convergence, i.e., $f_j \to f$ in $\mathcal{S}'$ if and only if $\langle f_j, \varphi \rangle \to \langle f, \varphi \rangle$ for all $\varphi \in \mathcal{S}$. For fundamentals on distributions, see [Ru3].

The Schwartz class $\mathcal{S}$ given in the above definition overlaps with the usual one by virtue of the lemma essentially due to Lemarié-Rieusset, see [LR].

LEMMA 3.2. *Suppose $\rho$ is a homogeneous quasi-norm associated with dilation $A$. Then*

$$(3.2) \qquad 1/c' \rho(x)^{\ln \lambda_- / \ln b} \leq |x| \leq c' \rho(x)^{\ln \lambda_+ / \ln b} \qquad \text{for } \rho(x) \geq 1,$$

$$(3.3) \qquad 1/c' \rho(x)^{\ln \lambda_+ / \ln b} \leq |x| \leq c' \rho(x)^{\ln \lambda_- / \ln b} \qquad \text{for } \rho(x) \leq 1,$$

*where $c'$ is some constant, and $\lambda_-$, $\lambda_+$ satisfy (2.1) and (2.2).*

PROOF. By Lemma 2.4, without loss of generality we can assume $\rho$ is the step homogeneous quasi-norm. For every integer $j \geq 0$ we have by (2.1)

$$\sup_{x \in B_{j+1} \setminus B_j} |x| = \sup_{x \in B_{j+1} \setminus B_j} |A^j A^{-j} x| \leq c \sup_{x \in B_1 \setminus B_0} |x| \lambda_+^j = c \sup_{x \in B_1 \setminus B_0} |x| \, b^{j \ln \lambda_+ / \ln b},$$

and analogously

$$\inf_{x \in B_{j+1} \setminus B_j} |x| \geq 1/c \inf_{x \in B_1 \setminus B_0} |x| \lambda_-^j = 1/c \inf_{x \in B_1 \setminus B_0} |x| \, b^{j \ln \lambda_- / \ln b}.$$

Therefore, (3.2) holds for $x \in B_{j+1} \setminus B_j$, where $j \geq 0$, so for all $x \in (B_0)^c$.

Considering $x \in B_{j+1} \setminus B_j$, where $j \leq 0$ and using (2.2) we obtain (3.3) for $x \in B_1$. $\square$

An important role in our investigation is played by the unit ball with respect to a particular finite family of pseudonorms of $\mathcal{S}$. It is critical that we use (3.1) to alter the standard definition of the pseudonorms in $\mathcal{S}$. Otherwise the grand maximal function introduced below would not behave nicely with respect to dilations and many results in Section 5 could not hold. A posteriori, we see that this is only a technical issue by virtue of Lemma 3.2.

DEFINITION 3.3. For an integer $N \geq 0$ consider family

$$(3.4) \qquad \mathcal{S}_N := \{\varphi \in \mathcal{S} : ||\varphi||_{\alpha,m} \leq 1 \text{ for } |\alpha| \leq N, m \leq N\}.$$

Equivalently

$$(3.5) \qquad \varphi \in \mathcal{S}_N \iff ||\varphi||_{\mathcal{S}_N} := \sup_{x \in \mathbb{R}^n} \sup_{|\alpha| \leq N} \max(1, \rho(x)^N) |\partial^\alpha \varphi(x)| \leq 1.$$

For $\varphi \in \mathcal{S}$, $k \in \mathbb{Z}$ define the dilate of $\varphi$ to the scale $k$ by

(3.6) $$\varphi_k(x) = b^{-k}\varphi(A^{-k}x).$$

DEFINITION 3.4. Suppose $\varphi \in \mathcal{S}$, and $f \in \mathcal{S}'$. The *nontangential maximal function* of $f$ with respect to $\varphi$ is defined as

(3.7) $$M_\varphi f(x) := \sup\{|f * \varphi_k(y)| : x - y \in B_k, k \in \mathbb{Z}\}.$$

The *radial maximal function* of $f$ with respect to $\varphi$ is defined as

(3.8) $$M_\varphi^0 f(x) := \sup_{k \in \mathbb{Z}} |f * \varphi_k(x)|.$$

For given $N \in \mathbb{N}$ we define the *nontangential grand maximal function* of $f$ as

(3.9) $$M_N f(x) := \sup_{\varphi \in \mathcal{S}_N} M_\varphi f(x).$$

The *radial grand maximal function* of $f$ is

(3.10) $$M_N^0 f(x) := \sup_{\varphi \in \mathcal{S}_N} M_\varphi^0 f(x).$$

Finally, given $\tilde{N} > 0$ we define the *tangential maximal function* of $f$ with respect to $\varphi$ as

$$T_\varphi^{\tilde{N}} f(x) = \sup\{|f * \varphi_k(y)|/\max(1, \rho(A^{-k}(x-y)))^{\tilde{N}} : y \in \mathbb{R}^n, k \in \mathbb{Z}\}.$$

REMARK. It is immediate that we have the following pointwise estimates between radial, nontangential, and tangential maximal functions

$$M_\varphi^0 f(x) \le M_\varphi f(x) \le T_\varphi^{\tilde{N}} f(x) \qquad \text{for all } x \in \mathbb{R}^n.$$

Moreover, radial and nontangential grand maximal functions are pointwise equivalent by virtue of Proposition 3.10. In Section 7, we will also see that for sufficiently large $N$ (depending on $\tilde{N}$) and $\varphi \in \mathcal{S}$ with $\int \varphi \ne 0$, the tangential maximal function $T_\varphi^{\tilde{N}} f(x)$ dominates pointwise the grand maximal function $M_N f(x)$, see Lemma 7.5.

PROPOSITION 3.5. *For $f \in \mathcal{S}'$, let $Mf$ denote any of the maximal functions introduced in Definition 3.4. Then $Mf : \mathbb{R}^n \to [0, \infty]$ is lower semicontinuous, function, i.e., for all $\lambda > 0$, $\{x \in \mathbb{R}^n : Mf(x) > \lambda\}$ is open.*

PROOF. If $\varphi \in \mathcal{S}$ and $f \in \mathcal{S}'$ then $f * \varphi$ is a continuous (even $C^\infty$) function on $\mathbb{R}^n$. Note that

$$M_\varphi f(x) = \sup\{|f * \varphi_k(y)| : x - y \in B_k \cap \mathbb{Q}^n, k \in \mathbb{Z}\}.$$

Furthermore, since the Schwartz class $\mathcal{S}$ is separable with respect to pseudonorms $\|\cdot\|_{\alpha,m}$, we can substitute $\mathcal{S}_N$ by a countable, dense subset in the definition of the grand maximal function. Therefore, in each case $Mf$ can be computed as a supremum of a countable family of continuous functions. Therefore, $Mf$ is lower semicontinuous. $\square$

We start the investigation of maximal functions with the fundamental Maximal Theorem 3.6. This and the following results are relatively mechanical conversions of the well-known classical results which also hold for homogeneous groups, see [FoS].

3. THE GRAND MAXIMAL DEFINITION OF ANISOTROPIC HARDY SPACES

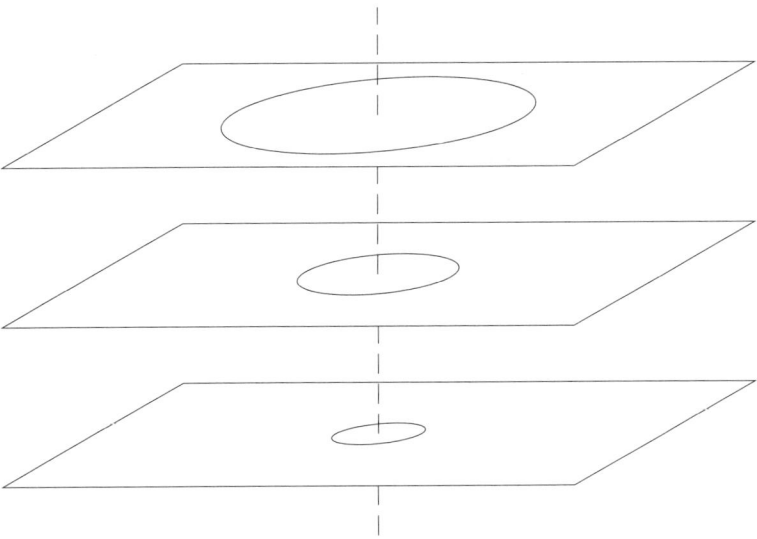

FIGURE 2. THE NONTANGENTIAL MAXIMAL FUNCTION $M_\varphi f(x)$ IS COMPUTED BY TAKING SUPREMA OF $|f * \varphi_k(y)|$ OVER ELLIPSES $x + B_k$ AT ALL SCALES $k \in \mathbb{Z}$.

The biggest novelty is the use of discrete (instead of continuous) dilations of fairly general form.

THEOREM 3.6 (The Maximal Theorem). *For a fixed $s > 1$ consider family*

$$(3.11) \qquad \mathcal{F} := \{\varphi \in L^\infty(\mathbb{R}^n) : |\varphi(x)| \leq (1 + \rho(x))^{-s}\}.$$

*For $1 \leq p \leq \infty$ and $f \in L^p(\mathbb{R}^n)$ define the maximal function*

$$(3.12) \qquad Mf(x) = M_\mathcal{F} f(x) := \sup_{\varphi \in \mathcal{F}} M_\varphi f(x).$$

*Then there exists a constant $C = C(s)$ so that*

$$(3.13) \quad |\{x : Mf(x) > \lambda\}| \leq C\|f\|_1/\lambda \qquad \text{for all } f \in L^1(\mathbb{R}^n),\ \lambda > 0,$$
$$(3.14) \quad \|Mf\|_p \leq Cp/(p-1)\|f\|_p \qquad \text{for all } f \in L^p(\mathbb{R}^n),\ 1 < p \leq \infty.$$

PROOF. For $\lambda, R > 0$ consider $\Omega_\lambda^R = \{x : Mf(x) > \lambda, |x| < R\}$. For every $x \in \Omega_\lambda^R$ take $y = y(x) \in \mathbb{R}^n$, $k = k(x) \in \mathbb{Z}$, and $\varphi \in \mathcal{F}$ such that $x - y \in B_k$ and $|f * \varphi_k(y)| > \lambda$.

$$\lambda < |f * \varphi_k(y)| \leq \int_{\mathbb{R}^n} |f(z)||\varphi_k(z-y)|dz = b^{-k}\int_{y+B_k}|f(z)||\varphi(A^{-k}(z-y))|dz$$
$$+ b^{-k}\sum_{j=1}^\infty \int_{y+B_{k+j}\setminus B_{k+j-1}} |f(z)||\varphi(A^{-k}(z-y))|dz$$
$$\leq b^{-k}\left[\sup_{z \in B_0}|\varphi(z)|\int_{y+B_k}|f(z)|dz + \sum_{j=1}^\infty \sup_{z \in B_{j+1}\setminus B_j}|\varphi(z)|\int_{y+B_{k+j}}|f(z)|dz\right]$$

$$\leq \left[1+\sum_{j=1}^{\infty} b^j \sup_{z \in B_{j+1} \setminus B_j} |\rho(z)|^{-s}\right] \sup_{j \geq 0} b^{-k-j} \int_{y+B_{k+j}} |f(z)| dz$$

$$\leq \sum_{j=0}^{\infty} b^{j(1-s)} \sup_{j \geq 0} b^{-k-j} \int_{y+B_{k+j}} |f(z)| dz.$$

Thus, there is $j_0 \geq 0$ so that

$$\frac{1}{b^{k+j_0}} \int_{y+B_{k+j_0}} |f(z)| dz > \lambda/C,$$

where $C = \sum_{j=0}^{\infty} b^{j(1-s)}$. Since $x - y \in B_k$,

$$y + B_{k+j_0} \subset x + B_k + B_{k+j_0} \subset x + 2B_{k+j_0} \subset x + B_{k+j_0+\omega}.$$

So for each $x \in \Omega_\lambda^R$ there is $r = r(x) = k + j_0 + \omega$, such that

$$\int_{x+B_r} |f(z)| dz > \frac{\lambda}{Cb^\omega} |B_r|.$$

By the Wiener Lemma there is sequence $(x_j)$ in $\Omega_\lambda^R$ so that $x_j + B_{r(x_j)}$ are mutually disjoint and $x_j + B_{r(x_j)+\omega}$ cover $\Omega_\lambda^R$. Therefore,

$$|\Omega_\lambda^R| \leq \sum_j |B_{r(x_j)+\omega}| \leq \sum_j b^\omega |B_{r(x_j)}| \leq \frac{Cb^{2\omega}}{\lambda} \|f\|_1.$$

By letting $R \to \infty$ we obtain (3.13).

Since $C = \int_{\mathbb{R}^n} (1+\rho(x))^{-s} dx < \infty$, we have $\|Mf\|_\infty \leq C\|f\|_\infty$ for $f \in L^\infty$. By the Marcinkiewicz Interpolation Theorem we obtain (3.14). $\square$

As a consequence of Theorem 3.6 we conclude that the *Hardy-Littlewood maximal function* $M = M_{HL}$ given by

$$(3.15) \qquad M_{HL}f(x) := \sup_{k \in \mathbb{Z}} \sup_{y \in x+B_k} \frac{1}{|B_k|} \int_{y+B_k} |f(z)| dz$$

satisfies (3.13) and (3.14). Naturally, this is also a consequence of general results for spaces of homogeneous type, see [St2, Theorem 1 in Chapter 1].

THEOREM 3.7. *Suppose $\varphi \in L^\infty(\mathbb{R}^n)$ satisfies $|\varphi(x)| \leq C(1+\rho(x))^{-s}$ for some $s > 1$, and let $c = \int \varphi$. Then for $f \in L^p(\mathbb{R}^n)$, $1 \leq p \leq \infty$,*

$$(3.16) \qquad \lim_{\substack{k \to -\infty \\ y \in x+B_k}} f * \varphi_k(y) = cf(x) \qquad \text{for a.e. } x \in \mathbb{R}^n.$$

PROOF. Suppose $p = 1$ and $f \in L^1(\mathbb{R}^n)$. By the Luzin (Лузин) Theorem given $\varepsilon > 0$ we can find a continuous function $g$ with compact support such that $\|f - g\|_1 < \varepsilon$. Clearly

$$\lim_{\substack{k \to -\infty \\ y \in x+B_k}} g * \varphi_k(y) = cg(x) \qquad \text{for all } x \in \mathbb{R}^n.$$

Since
$$\limsup_{\substack{k\to -\infty \\ y\in x+B_k}} |f*\varphi_k(y) - cf(x)| = \limsup_{\substack{k\to -\infty \\ y\in x+B_k}} |f*\varphi_k(y) - g*\varphi_k(x) + g*\varphi_k(x) - cf(x)|$$
$$\leq \sup_{k\in\mathbb{Z}} \sup_{y\in x+B_k} |f*\varphi_k(y) - g*\varphi_k(x)| + |c||g(x) - f(x)|$$
$$\leq M_\varphi(f-g)(x) + |c||f(x) - g(x)|,$$

by Theorem 3.6 and the Chebyshev (Чебышев) inequality we have for any $\lambda > 0$

$$|\{x : \limsup_{\substack{k\to -\infty \\ y\in x+B_k}} |f*\varphi_k(y) - cf(x)| > \lambda\}|$$
$$\leq |\{x : M_\varphi(f-g)(x) > \lambda/2\}| + |\{x : |c||f(x) - g(x)| > \lambda/2\}|$$
$$\leq C\|f-g\|_1/\lambda < C\varepsilon/\lambda.$$

Since $\lambda > 0$ is arbitrary we have (3.16).

Suppose $p > 1$ and $f \in L^p(\mathbb{R}^n)$. Given $j \in \mathbb{Z}$ let $g = f\mathbf{1}_{B_j}$. Since $g \in L^1(\mathbb{R}^n)$,

$$\lim_{\substack{k\to -\infty \\ y\in x+B_k}} g*\varphi_k(y) = cg(x) = cf(x) \qquad \text{for a.e. } x \in B_j.$$

If $x \in B_j$ we can choose $K \in \mathbb{Z}$ so that $x + B_{K+\omega} \subset B_j$. If $y \in x+B_k$ for some $k \leq K$ and $z \in (B_j)^c \subset x + (B_{K+\omega})^c$ then $y - z \in (B_K)^c$ by (2.11). Hence by Hölder's inequality, $1/p + 1/q = 1$,

$$|(f-g)*\varphi_k(y)| = \left|\int_{(B_j)^c} f(z)\varphi_k(y-z)dz\right| \leq \|f\|_p \left(\int_{(B_j)^c} |\varphi_k(y-z)|^q dz\right)^{1/q}$$
$$\leq \|f\|_p \left(\int_{(B_K)^c} |\varphi_k(z)|^q dz\right)^{1/q} \leq \|f\|_p b^{-k+k/q} \left(\int_{(B_{K-k})^c} |\varphi(z)|^q dz\right)^{1/q}$$
$$\leq b^{-k+k/q}\|f\|_p \left(\int_{(B_{K-k})^c} |\rho(z)|^{-qs} dz\right)^{1/q} \leq Cb^{-k+k/q} b^{(K-k)(-s+1/q)}$$
$$= Cb^{K(-s+1/q)} b^{k(s-1)} \to 0 \qquad \text{as } k \to -\infty.$$

Therefore, (3.16) holds for a.e. $x \in B_j$ and since $j$ is arbitrary it holds for a.e. $x \in \mathbb{R}^n$. □

The next lemma is a basic approximation of identity result for the space of tempered distributions $\mathcal{S}'$. Since it is a more general variant of the result common in the literature we include its proof.

LEMMA 3.8. *Suppose $\varphi \in \mathcal{S}$ and $\int \varphi = 1$. Then for any $\psi \in \mathcal{S}$ and $f \in \mathcal{S}'$ we have*

(3.17) $\qquad\qquad\qquad \psi * \varphi_k \to \psi \quad \text{in } \mathcal{S} \text{ as } k \to -\infty,$
(3.18) $\qquad\qquad\qquad f * \varphi_k \to f \quad \text{in } \mathcal{S}' \text{ as } k \to -\infty.$

In Lemma 3.8 we can relax the condition that $\varphi_k$'s are given by (3.6). Instead, we only need to assume that $\varphi_k(x) = |\det A_k|\varphi(A_k x)$ for some nondegenerate matrices $A_k$ such that $\|A_k^{-1}\| \to 0$ as $k \to -\infty$.

PROOF. It is clear that $\psi * \varphi_k(x) \to \psi(x)$ pointwise as $k \to -\infty$. Since $\partial^\alpha(\psi * \varphi_k)(x) = (\partial^\alpha \psi) * \varphi_k(x) \to \partial^\alpha \psi(x)$ pointwise as $k \to -\infty$ for every multi-index $\alpha$, it suffices to show

$$\sup_{x \in \mathbb{R}^n} |x|^N |\psi * \varphi_k(x) - \psi(x)| \to 0 \quad \text{as } k \to -\infty, \tag{3.19}$$

for every integer $N \geq 0$.

Given $\varepsilon > 0$ find $\delta > 0$ so that $|x|^N |\psi(y-x) - \psi(x)| < \varepsilon$ for all $|y| \leq \delta$. Let $K$ be such that

$$\int_{|y|>\delta} (1+|y|)^N |\varphi_k(y)| dy < \varepsilon \quad \text{for } k \leq K.$$

If $k \leq K$

$$|x|^N |\psi * \varphi_k(x) - \psi(x)| = \left| \int_{\mathbb{R}^n} |x|^N (\psi(y-x) - \psi(x)) \varphi_k(y) dy \right|$$

$$\leq \int_{|y|\leq\delta} |x|^N |\psi(y-x) - \psi(x)| |\varphi_k(y)| dy + \int_{|y|>\delta} |x|^N |\psi(y-x) - \psi(x)| |\varphi_k(y)| dy$$

$$\leq \varepsilon \int_{\mathbb{R}^n} |\varphi_k(y)| dy + \sup_{x \in \mathbb{R}^n} |x|^N |\psi(x)|^N \int_{|y|>\delta} |\varphi_k(y)| dy$$

$$+ \int_{|y|>\delta} C|x|^N (1+|x-y|)^{-N} (1+|y|)^{-N} (1+|y|)^N |\varphi_k(y)| dy$$

$$\leq C'\varepsilon + C' \int_{|y|>\delta} (1+|y|)^N |\varphi_k(y)| dy \leq 2C'\varepsilon.$$

This shows (3.19) and thus (3.17). (3.18) follows from (3.17) since $\langle f * \varphi_k, \psi \rangle = \langle f, \psi * \tilde{\varphi}_k \rangle$, where $\tilde{\varphi}_k(x) = \varphi_k(-x)$. □

The next theorem can be thought as a converse to the Maximal Theorem 3.6 for $p > 1$.

THEOREM 3.9. *Suppose $f \in \mathcal{S}'$, $\varphi \in \mathcal{S}$, $\int \varphi \neq 0$, and $1 \leq p \leq \infty$. If $M_\varphi^0 f \in L^p(\mathbb{R}^n)$ then $f \in L^p(\mathbb{R}^n)$.*

PROOF. Without loss of generality we can assume that $\int \varphi = 1$. Suppose $p > 1$. Since the set $\{f * \varphi_k : k \in \mathbb{Z}\}$ is bounded in $L^p$, by the Alaoglu Theorem there is a sequence $k_j \to -\infty$ such that $f * \varphi_{k_j}$ converges weak-* in $L^p$, and hence in $\mathcal{S}'$. By Lemma 3.8 this limit is $f$ and thus $f \in L^p$.

Suppose now $p = 1$. By [Wo1, Theorem III.C.12] the set $\{f * \varphi_k : k \in \mathbb{Z}\}$ is relatively weakly compact in $L^1$ and by the Eberlein-Šmulian (Шмулян) Theorem there is a sequence $k_j \to -\infty$ such that $f * \varphi_{k_j}$ converges weakly in $L^1$, and hence in $\mathcal{S}'$. By Lemma 3.8 this limit is $f$ and thus $f \in L^1$. Alternatively, we could think of the set $\{f * \varphi_k : k \in \mathbb{Z}\}$ as a bounded set in the space of finite complex Borel measures on $\mathbb{R}^n$ and use Alaoglu Theorem to find a sequence $k_j \to -\infty$ such that $f * \varphi_{k_j}$ converges weak-* to an absolutely continuous measure. This measure is $f(x) dx$ by Lemma 3.8. □

REMARK. Theorems 3.6 and 3.9 together assert that for $p > 1$ the subclass of regular $L^p$ integrable distributions in $\mathcal{S}'$ is invariant with respect to maximal functions introduced in Definition 3.4. Indeed, it follows from Theorems 3.6 and 3.9 that for $1 < p \leq \infty$ the following are equivalent for a distribution $f \in \mathcal{S}'$,

- $f$ is regular and belongs to $L^p$,
- $M_\varphi^0 f \in L^p$ (or $M_\varphi f \in L^p$) for every $\varphi \in \mathcal{S}$,
- $M_\varphi^0 f \in L^p$ (or $M_\varphi f \in L^p$) for some $\varphi$ with $\int \varphi \neq 0$,
- $M_N^0 f \in L^p$ for some (or every) $N \geq 2$.

The next result asserts that radial and nontangential grand maximal functions are pointwise equivalent.

PROPOSITION 3.10. *For every $N \geq 0$ there is a constant $C = C(N)$ so that for all $f \in \mathcal{S}'$,*

$$(3.20) \qquad M_N^0 f(x) \leq M_N f(x) \leq C M_N^0 f(x) \qquad \text{for } x \in \mathbb{R}^n.$$

PROOF. The first inequality is obvious. To see the second inequality, note that

$$\begin{aligned}
M_N f(x) &= \sup\{|(f * \varphi_k)(x + A^k y)| : k \in \mathbb{Z}, y \in B_0, \varphi \in \mathcal{S}_N\} \\
(3.21) &= \sup\{|(f * \phi_k)(x)| : k \in \mathbb{Z}, \phi(z) = \varphi(z + y) \text{ for some } y \in B_0, \varphi \in \mathcal{S}_N\} \\
&= \sup\{M_\phi^0(x) : \phi(z) = \varphi(z + y) \text{ for some } y \in B_0, \varphi \in \mathcal{S}_N\}.
\end{aligned}$$

By (2.12) if $\phi(x) = \varphi(x + y)$ for some $y \in B_0$ then

$$(3.22) \qquad \begin{aligned}
\|\phi\|_{\mathcal{S}_N} &= \sup_{x \in \mathbb{R}^n} \sup_{|\alpha| \leq N} \max(1, \rho(x)^N) |\partial^\alpha \varphi(x + y)| \\
&= \sup_{x \in \mathbb{R}^n} \sup_{|\alpha| \leq N} \max(1, \rho(x - y)^N) |\partial^\alpha \varphi(x)| \\
&\leq b^{\omega N} \sup_{x \in \mathbb{R}^n} \sup_{|\alpha| \leq N} \max(1, \rho(x)^N) |\partial^\alpha \varphi(x)| = b^{\omega N} \|\varphi\|_{\mathcal{S}_N}.
\end{aligned}$$

Combining (3.21) and (3.22) we have

$$M_N f(x) \leq \sup\{M_\phi^0 f(x) : \phi \in \mathcal{S}, \|\phi\|_{\mathcal{S}_N} \leq b^{\omega N}\} \leq b^{\omega N} M_N^0 f(x),$$

which shows (3.20). □

We are now ready to state the definition of anisotropic Hardy spaces.

DEFINITION 3.11. For a given dilation $A$ and $0 < p < \infty$ we denote

$$N_p := \begin{cases} \lfloor (1/p - 1) \ln b / \ln \lambda_- \rfloor + 2 & 0 < p \leq 1, \\ 2 & p > 1. \end{cases}$$

For every $N \geq N_p$ we define the *anisotropic Hardy space* associated with the dilation $A$ as

$$(3.23) \qquad H^p(\mathbb{R}^n) = H_A^p(\mathbb{R}^n) = \{f \in \mathcal{S}' : M_N f \in L^p\},$$

with the quasi-norm $\|f\|_{H^p} = \|M_N f\|_p$.

Since the dimension $n$ and the dilation $A$ remain constant throughout this chapter (except Section 10) we are going to denote the anisotropic Hardy space by $H^p$ or $H^p_{(N)}$. Even though quasi-norms on $H^p$ depend on the choice of $N$, it follows from Theorems 4.2 and 6.4 that the above definition of $H^p$ does not depend on the choice of $N$ as long as $N \geq N_p$. To escape possible ambiguity and to fix the attention the reader can think that $N = N_p$ in (3.23).

By the above Remark we have $H^p = L^p$ for $p > 1$ irrespective of the dilation $A$. Moreover, by Theorem 3.9 we have $H^1 \subset L^1$. Therefore, only for $0 < p \leq 1$ do anisotropic Hardy spaces $H^p$ merit further investigation.

PROPOSITION 3.12. *The space $H^p$ is complete.*

PROOF. Since $H^p = L^p$ for $p > 1$ we only need to consider $p \leq 1$.

For every $\varphi \in \mathcal{S}$ and every sequence $(f_i)_{i \in \mathbb{N}}$ in $\mathcal{S}'$ such that $\sum_i f_i$ converges in $\mathcal{S}'$ to the tempered distribution $f$, the series $\sum_i f_i * \varphi(x)$ converges pointwise to $f * \varphi(x)$ for each $x \in \mathbb{R}^n$. Thus,

$$M_N f(x)^p \leq \left(\sum_i M_N f_i(x)\right)^p \leq \sum_i (M_N f_i(x))^p \qquad \text{for all } x \in \mathbb{R}^n,$$

and we have $\|f\|_{H^p}^p \leq \sum_i \|f_i\|_{H^p}^p$.

To prove that $H^p$ is complete, it suffices to show that for every sequence $(f_i)$ such that $\|f_i\|_{H^p} < 2^{-i}$ for $i \in \mathbb{N}$, the series $\sum_i f_i$ converges in $H^p$. Since partial sums $\sum_i f_i$ are Cauchy in $H^p$, hence they are Cauchy in $\mathcal{S}'$, and $\sum_i f_i$ converges in $\mathcal{S}'$ to some $f$, because $\mathcal{S}'$ is complete. Therefore,

$$\|f - \sum_{i=1}^{j} f_i\|_{H^p}^p = \|\sum_{i=j+1}^{\infty} f_i\|_{H^p}^p \leq \sum_{i=j+1}^{\infty} \|f_i\|_{H^p}^p \leq \sum_{i=j+1}^{\infty} 2^{-ip} \to 0 \quad \text{as } j \to \infty.$$

This finishes the proof. □

## 4. The atomic definition of anisotropic Hardy spaces

DEFINITION 4.1. We say a triplet $(p, q, s)$ is *admissible* (with respect to the dilation $A$) if $0 < p \leq 1$, $1 \leq q \leq \infty$, $p < q$, $s \in \mathbb{N}$, and $s \geq \lfloor (1/p - 1) \ln b / \ln \lambda_- \rfloor$. A $(p, q, s)$-atom (associated with the dilation $A$) is a function $a$ such that

(4.1) $\quad\quad\quad\quad \operatorname{supp} a \subset B_j + x_0 \quad$ for some $j \in \mathbb{Z}, x_0 \in \mathbb{R}^n$,

(4.2) $\quad\quad\quad\quad \|a\|_q \leq |B_j|^{1/q - 1/p}$,

(4.3) $\quad\quad\quad\quad \int_{\mathbb{R}^n} a(x) x^\alpha dx = 0 \quad$ for $|\alpha| \leq s$.

The rationale behind the conditions imposed on atoms is revealed in the next theorem.

THEOREM 4.2. *Suppose $(p, q, s)$ is admissible and $N \geq N_p$. Then there is a constant $\tilde{C}$ depending only on $p$ and $q$ such that for all $(p, q, s)$-atoms $a$ we have $\|M_N a\|_p \leq \tilde{C}$.*

PROOF. Since $\mathcal{S}_N \subset \mathcal{S}_{N_p}$, we have $M_N a(x) \leq M_{N_p} a(x)$, and we can assume that $N = N_p$. By Proposition 3.10 it suffices to show $\|Ma\|_p \leq \tilde{C}$, where $M = M_N^0$ is the radial grand maximal function. It also suffices to consider only $(p, q, s)$-atoms $a$ with minimal value of $s = \lfloor (1/p - 1) \ln b / \ln \lambda_- \rfloor$.

Suppose an atom $a$ is associated with the ball $x_0 + B_j$ for some $x_0 \in \mathbb{R}^n$, $j \in \mathbb{Z}$. We estimate separately on $x_0 + B_{j+\omega}$ and $(x_0 + B_{j+\omega})^c$.

**Case I.** On $B_{j+\omega}$ we use the Maximal Theorem 3.6 for $q > 1$ and Hölder's inequality

$$\int_{x+B_{j+\omega}} Ma(x)^p dx \leq \left( \int_{x+B_{j+\omega}} Ma(x)^q dx \right)^{p/q} |B_{j+\omega}|^{1-p/q}$$
$$\leq C \|a\|_q^p |B_j|^{1-p/q} \leq C.$$

If $q = 1$, take $\lambda > 0$, and consider $\Omega_\lambda = \{x : Ma(x) > \lambda\}$. Then

$$|\Omega_\lambda \cap B_{j+\omega}| \leq \min(|\Omega_\lambda|, |B_{j+\omega}|) \leq \min(C\|a\|_1/\lambda, |B_{j+\omega}|)$$
$$= \min(C|B_j|^{1-1/p}/\lambda, b^\omega |B_j|) \leq C \min(|B_j|^{1-1/p}/\lambda, |B_j|).$$

We have equality between the two terms in the last minimum if $\lambda = |B_j|^{-1/p}$. Therefore,

$$\int_{B_{j+\omega}} Ma(x)^p dx = \int_0^\infty p\lambda^{p-1} |\Omega_\lambda \cap B_{j+\omega}| d\lambda$$
$$\leq C \int_0^{|B_j|^{-1/p}} |B_j| p\lambda^{p-1} d\lambda + \int_{|B_j|^{-1/p}}^\infty |B_j|^{1-1/p} p\lambda^{p-2} d\lambda = C.$$

**Case II.** Take $x \in (x_0 + B_{j+\omega})^c$, $\varphi \in \mathcal{S}_N$, and $k \in \mathbb{Z}$. Suppose $P$ is a polynomial of degree $\leq s$ to be specified later. Then we have

$$|(a * \varphi_k)(x)| = b^{-k} \left| \int_{\mathbb{R}^n} a(y) \varphi(A^{-k}(x-y)) dy \right|$$
$$= b^{-k} \left| \int_{x_0+B_j} a(y)(\varphi(A^{-k}(x-y)) - P(A^{-k}(x-y))) dy \right|$$

$$\leq b^{-k}||a||_q \left( \int_{x_0+B_j} |\varphi(A^{-k}(x-y)) - P(A^{-k}(x-y))|^{q'} dy \right)^{1/q'}$$

(4.4)
$$\leq b^{-k} b^{j(1/q-1/p)} b^{k/q'} \left( \int_{A^{-k}(x-x_0)+B_{j-k}} |\varphi(y) - P(y)|^{q'} dy \right)^{1/q'}$$

$$\leq b^{-k} b^{j(1/q-1/p)} b^{k/q'} b^{(j-k)/q'} \sup_{y \in A^{-k}(x-x_0)+B_{j-k}} |\varphi(y) - P(y)|$$

$$= b^{-j/p} b^{j-k} \sup_{y \in A^{-k}(x-x_0)+B_{j-k}} |\varphi(y) - P(y)|,$$

where $q'$ denotes the conjugate power to $q$, $1/q + 1/q' = 1$. Even though the above calculation holds for $q > 1$ it can be easily adjusted for the case $q = 1$ to yield the same estimate.

Suppose $x \in x_0 + B_{j+\omega+m+1} \setminus B_{j+\omega+m}$ for some integer $m \geq 0$. Then $A^{-k}(x - x_0) + B_{j-k} \subset A^{-k}(B_{j+\omega+m+1} \setminus B_{j+\omega+m}) + B_{j-k} = A^{j-k}((B_{\omega+m+1} \setminus B_{\omega+m}) + B_0) \subset A^{j-k}(B_m)^c$ by virtue of (2.11). We consider two cases. If $j \geq k$ then we choose polynomial $P = 0$, and

(4.5) $$\sup_{y \in A^{-k}(x-x_0)+B_{j-k}} |\varphi(y)| \leq \sup_{y \in A^{-k}(x-x_0)+B_{j-k}} \min(1, \rho(y)^{-N}) \leq b^{-N(j-k+m)}.$$

If $j < k$ then we choose polynomial $P$ to be the Taylor expansion of $\varphi$ at the point $A^{-k}(x - x_0)$ of order $s$. By the Taylor Remainder Theorem and (2.2) we have

(4.6)
$$\sup_{y \in A^{-k}(x-x_0)+B_{j-k}} |\varphi(y) - P(y)|$$
$$\leq C \sup_{z \in B_{j-k}} \sup_{|\alpha|=s+1} |\partial^\alpha \varphi(A^{-k}(x-x_0)+z)||z|^{s+1}$$
$$\leq C \lambda_-^{(s+1)(j-k)} \sup_{y \in A^{-k}(x-x_0)+B_{j-k}} \min(1, \rho(y)^{-N})$$
$$\leq C \lambda_-^{(s+1)(j-k)} \min(1, b^{-N(j-k+m)}).$$

Combining (4.4), (4.5) and (4.6) we have for $x \in x_0 + B_{j+\omega+m+1} \setminus B_{j+\omega+m}$, $m \geq 0$,

$$Ma(x)^p = \sup_{\varphi \in \mathcal{S}_N} \sup_{k \in \mathbb{Z}} |(a * \varphi_k)(x)|^p \leq b^{-j} \max \left( \sup_{k \in \mathbb{Z}, k \leq j} b^{p(j-k)} b^{-Np(j-k+m)} \right.$$
$$\left. + C \sup_{k \in \mathbb{Z}, k > j} b^{p(j-k)} \lambda_-^{p(s+1)(j-k)} \min(1, b^{-Np(j-k+m)}) \right).$$

Note that the supremum over $k \leq j$ has the largest value for $k = j$. Since $N \geq s+2$ we have $b\lambda_-^{s+1} \leq b^N$, and the supremum over $k > j$ is attained when $j - k + m = 0$. Indeed, it suffices to check for the maximum value in the range $j < k \leq j + m$ and $k \geq j + m$. Therefore,

$$Ma(x)^p \leq b^{-j} \max(b^{-Npm}, C(b\lambda_-^{s+1})^{-pm}) \leq Cb^{-j}(b\lambda_-^{s+1})^{-pm},$$

again by $b\lambda_-^{s+1}b^{-N} \leq 1$. Since $s = \lfloor(1/p - 1)\ln b/\ln\lambda_-\rfloor$, we have $s > (1/p - 1)\ln b/\ln\lambda_- - 1$, and thus $\lambda_-^{s+1}b^{1-1/p} > 1$. Therefore,

$$\int_{(x_0+B_j)^c} Ma(x)^p dx = \sum_{m=0}^{\infty} |B_{j+\omega+m+1} \setminus B_{j+\omega+m}| \sup_{x \in x_0+B_{j+\omega+m+1}\setminus B_{j+\omega+m}} Ma(x)^p$$

$$\leq C \sum_{m=0}^{\infty} b^{j+\omega+m+1}b^{-j}(b\lambda_-^{s+1})^{-pm} = Cb^{\omega+1}\sum_{m=0}^{\infty}(b^{p-1}\lambda_-^{(s+1)p})^{-m} = C' < \infty.$$

Combining case I and II and we see that $||Ma||_p \leq \tilde{C}$ for some constant $\tilde{C}$ independent of the choice of a $(p,q,s)$-atom $a$. □

The proof of this result could be simplified by virtue of Theorem 7.1, since it is easier to estimate $M_\varphi^0 f$ than $M_N f$. We are now ready to introduce the anisotropic Hardy spaces in terms of atomic decompositions.

DEFINITION 4.3. For a given dilation $A$ and an admissible triplet $(p,q,s)$ we define the *atomic anisotropic Hardy space* $H_{q,s}^p$ associated with dilation $A$ as the set of all tempered distributions $f \in \mathcal{S}'$ of the form $\sum_{i=1}^{\infty} \kappa_i a_i$ with convergence in $\mathcal{S}'$, where $\sum_{i=1}^{\infty} |\kappa_i|^p < \infty$, and $a_i$ is a $(p,q,s)$-atom for each $i \in \mathbb{N}$. The quasi-norm of $f \in H_{q,s}^p$ is defined as

$$||f||_{H_{q,s}^p} = \inf\left\{\left(\sum_{i=1}^{\infty}|\kappa_i|^p\right)^{1/p} : f = \sum_{i=1}^{\infty}\kappa_i a_i, \quad a_i \text{ is a } (p,q,s)\text{-atom for } i \in \mathbb{N}\right\}.$$

It follows from Theorem 4.5 that the series $\sum_{i=1}^{\infty} \kappa_i a_i$ converges in $\mathcal{S}'$ for every choice of $\kappa_i$ and $(p,q,s)$-atoms $a_i$ with $\sum_{i=1}^{\infty}|\kappa_i|^p < \infty$. The representation of $f \in \mathcal{S}'$ in this form $\sum_{i=1}^{\infty}\kappa_i a_i$ is referred as an *atomic decomposition* of $f$. Furthermore, if $||f||_{H_{q,s}^p} = 0$ then necessarily $f = 0$.

PROPOSITION 4.4. *Suppose the triplet $(p,q,s)$ is admissible. Then the space $H_{q,s}^p$ is complete.*

The proof of Proposition 4.4 is a routine, see Proposition 3.12.

THEOREM 4.5. *If the triplet $(p,q,s)$ is admissible and $N \geq N_p$ then $H_{q,s}^p \subset H^p \subset \mathcal{S}'$, where $H^p$ is defined using maximal function $M_N$. Moreover, the inclusion maps are continuous.*

PROOF. Suppose $f \in H_{q,s}^p$ has an atomic decomposition $f = \sum_{i=1}^{\infty}\kappa_i a_i$. Then

$$||f||_{H^p}^p = \int_{\mathbb{R}^n}(M_N(\sum_{i=1}^{\infty}\kappa_i a_i)(x))^p dx \leq \sum_{i=1}^{\infty}|\kappa_i|^p\int_{\mathbb{R}^n}M_N a_i(x)^p dx \leq \tilde{C}\sum_{i=1}^{\infty}|\kappa_i|^p,$$

where $\tilde{C}$ is the constant in Theorem 4.2. Since we can choose $(\sum_{i=1}^{\infty}|\kappa_i|^p)^{1/p}$ to be arbitrarily close to $||f||_{H_{q,s}^p}$ thus $||f||_{H^p} \leq \tilde{C}^{1/p}||f||_{H_{q,s}^p}$, and the inclusion $H_{q,s}^p \hookrightarrow H^p$ is continuous.

Suppose next $f \in H^p$, and $\varphi \in \mathcal{S}$.

$$|\langle f, \varphi\rangle| = |f * \tilde{\varphi}(0)| \leq M_{\tilde{\varphi}}f(x) \quad \text{for } x \in B_0,$$

where $\tilde{\varphi}(x) = \varphi(-x)$. Therefore,

$$|\langle f, \varphi \rangle|^p = |B_0||\langle f, \varphi \rangle|^p \leq \int_{B_0} M_{\tilde{\varphi}} f(x)^p dx \leq \int_{\mathbb{R}^n} M_{\tilde{\varphi}} f(x)^p dx$$

$$\leq ||\tilde{\varphi}||^p_{\mathcal{S}_N} \int_{\mathbb{R}^n} M_N f(x)^p dx = ||\tilde{\varphi}||^p_{\mathcal{S}_N} ||f||^p_{H^p}.$$

If a sequence $(f_i)$ converges to $f$ in $H^p$, then $\langle f_i, \varphi \rangle \to \langle f, \varphi \rangle$ as $i \to \infty$, and the inclusion $H^p \hookrightarrow \mathcal{S}'$ is continuous. □

## 5. The Calderón-Zygmund decomposition for the grand maximal function

In this section we define and investigate extensively the Calderón-Zygmund decomposition. The excellent exposition of this decomposition in the setting of homogeneous groups is in [FoS]. Even though we work only in $\mathbb{R}^n$ we allow much more general dilation structures than in [FoS] and therefore we need to build the machinery from scratch. We try to follow as closely as possible the skeleton of the construction in [FoS] to give the reader familiar with this book a comfortable path through numerous technical nuances.

Throughout this section we consider a tempered distribution $f$ so that

$$|\{x : Mf(x) > \lambda\}| < \infty \qquad \text{for all } \lambda > 0,$$

where $M = M_N$ for some fixed integer $N \geq 2$. Later with regard to the Hardy space $H^p$ ($0 < p \leq 1$) we restrict attention to $N \geq N_p := \lfloor (1/p - 1) \ln b / \ln \lambda_- \rfloor + 2$. For a fixed $\lambda > 0$ we set $\Omega = \{x : Mf(x) > \lambda\}$. The Whitney Lemma applied to $\Omega$ with $d = 4\omega$ yields a sequence $(x_j)_{j \in \mathbb{N}} \subset \Omega$, and a sequence of integers $(l_j)_{j \in \mathbb{N}}$, so that

(5.1) $$\Omega = \bigcup_{j \in \mathbb{N}} (x_j + B_{l_j}),$$

(5.2) $$(x_i + B_{l_i - \omega}) \cap (x_j + B_{l_j - \omega}) = \emptyset \qquad \text{for } i \neq j,$$

(5.3) $$(x_j + B_{l_j + \gamma - 1}) \cap \Omega^c = \emptyset, \quad (x_j + B_{l_j + \gamma}) \cap \Omega^c \neq \emptyset \qquad \text{for } j \in \mathbb{N},$$

(5.4) $$(x_i + B_{l_i + 2\omega}) \cap (x_j + B_{l_j + 2\omega}) \neq \emptyset \implies |l_i - l_j| \leq \omega,$$

(5.5) $$\#\{j \in \mathbb{N} : (x_i + B_{l_i + 2\omega}) \cap (x_j + B_{l_j + 2\omega}) \neq \emptyset\} \leq L \qquad \text{for } i \in \mathbb{N},$$

where $\gamma := d + 1 = 4\omega + 1$, and $L$ is some constant independent of $\Omega$.

Fix $\theta \in C^\infty(\mathbb{R}^n)$ such that $\operatorname{supp} \theta \subset B_\omega$, $0 \leq \theta \leq 1$, and $\theta \equiv 1$ on $B_0$. For every $j \in \mathbb{N}$ define

(5.6) $$\theta_j(x) = \theta(A^{-l_j}(x - x_j)).$$

One should think of $\theta_j$ as the localized version of $\theta$ to the scale $l_j$ centered at $x_j$, and corresponding to the ball $x_j + B_{l_j}$ in the Whitney decomposition. Clearly $\operatorname{supp} \theta_j \subset x_j + B_{l_j + \omega}$, $\theta_j \equiv 1$ on $x_j + B_{l_j}$, and by (5.1) and (5.5)

(5.7) $$1 \leq \sum_{j \in \mathbb{N}} \theta_j(x) \leq L \qquad \text{for } x \in \Omega.$$

For every $i \in \mathbb{N}$ define

(5.8) $$\zeta_i(x) = \begin{cases} \theta_i(x) / \sum_j \theta_j(x) & x \in \Omega, \\ 0 & x \notin \Omega. \end{cases}$$

We have $\zeta_i \in C^\infty(\mathbb{R}^n)$, $\operatorname{supp} \zeta_i \subset x_i + B_{l_i + \omega}$, $0 \leq \zeta_i \leq 1$, $\zeta_i \equiv 1$ on $x_i + B_{l_i - \omega}$ by (5.2), and $\sum_{i \in \mathbb{N}} \zeta_i = \mathbf{1}_\Omega$. Therefore, the family $\{\zeta_i\}$ forms a smooth partition of unity which is subordinate to the covering of $\Omega$ by the balls $\{x_i + B_{l_i + \omega}\}$.

Let $\mathcal{P}_s$ denote the linear space of polynomials in $n$ variables of degree $\leq s$, where $s \geq 0$ is a fixed integer. For each $i \in \mathbb{N}$ we introduce the norm in the space $\mathcal{P}_s$ by setting

$$(5.9) \qquad \|P\| = \left( \frac{1}{\int \zeta_i} \int_{\mathbb{R}^n} |P(x)|^2 \zeta_i(x) dx \right)^{1/2} \qquad \text{for } P \in \mathcal{P}_s,$$

which makes $\mathcal{P}_s$ a finite dimensional Hilbert space. The distribution $f \in \mathcal{S}'$ induces a linear functional on $\mathcal{P}_s$ by

$$Q \mapsto \frac{1}{\int \zeta_i} \langle f, Q\zeta_i \rangle \qquad \text{for } Q \in \mathcal{P}_s,$$

which is represented by the Riesz Lemma by the unique polynomial $P_i \in \mathcal{P}_s$ such that

$$\frac{1}{\int \zeta_i} \langle f, Q\zeta_i \rangle = \frac{1}{\int \zeta_i} \langle P_i, Q\zeta_i \rangle = \frac{1}{\int \zeta_i} \int_{\mathbb{R}^n} P_i(x) Q(x) \zeta_i(x) dx \qquad \text{for all } Q \in \mathcal{P}_s.$$

For every $i \in \mathbb{N}$ define distributions $b_i := (f - P_i)\zeta_i$.

We will show that for a suitable choice of $s$ and $N$ the series $\sum_i b_i$ converges in $\mathcal{S}'$. Then we can define $g := f - \sum_i b_i$.

DEFINITION 5.1. The representation $f = g + \sum b_i$, where $g$ and $b_i$ are as above is a *Calderón-Zygmund decomposition* of degree $s$ and height $\lambda$ associated with $M_N f$.

Intuitively, one should think of $g$ as the good part of $f$ and of $b_i$'s as the bad parts of $f$. For a suitable choice of $s$ and $N$, the good part $g$ behaves very nicely. In particular, $g$ is in $L^1$ and, in addition, it is in $L^\infty$ whenever $f$ is in $L^1$, see Lemma 5.10. On the other hand, the bad parts $b_i$'s are well-localized on the balls coming from the Whitney decomposition; they have a certain number of vanishing moments, and thus they can be nicely controlled in terms of the grand maximal function of $f$, see Lemma 5.7.

The rest of this section consists of a series of lemmas. In Lemmas 5.2 and 5.3 we show properties of the smooth partition of unity $\{\zeta_i\}$. In Lemmas 5.4, 5.6, 5.7, and 5.8 we derive the estimates for the bad parts $b_i$'s using Lemma 5.5, which is a result about Taylor polynomials of independent interest. Lemmas 5.9 and 5.10 give controls for the good part $g$. Finally, the culmination of this section is Corollary 5.11 showing the density of $L^1 \cap H^p$ functions in $H^p$ spaces.

LEMMA 5.2. *There exists a constant $A_1 > 0$ depending only on $N$, so that for all $i \in \mathbb{N}$ and $l \leq l_i$*

$$(5.10) \qquad \sup_{|\alpha| \leq N} \sup_{x \in \mathbb{R}^n} |\partial^\alpha \tilde{\zeta}(x)| \leq A_1, \qquad \text{where } \tilde{\zeta}(x) := \zeta_i(A^l x).$$

PROOF. For a given $i \in \mathbb{N}$ consider $J = \{j \in \mathbb{N} : (x_i + B_{l_i + \omega}) \cap (x_j + B_{l_j + \omega}) \neq \emptyset\}$. Since $\operatorname{supp} \zeta_i \subset x_i + B_{l_i + \omega}$, thus $\operatorname{supp} \tilde{\zeta} \subset A^{-l} x_i + B_{l_i - l + \omega}$. Therefore, we need to consider only $x \in A^{-l} x_i + B_{l_i - l + \omega}$ in the supremum in (5.10). If $A^l x \in x_i + B_{l_i + \omega}$ then $\theta_j(A^l x) = 0$ for $j \in \mathbb{N} \setminus J$ and we have

$$\tilde{\zeta}(x) = \zeta_i(A^l x) = \frac{\theta_i(A^l x)}{\sum_{j \in \mathbb{N}} \theta_j(A^l x)} = \frac{\theta_i(A^l x)}{\sum_{j \in J} \theta_j(A^l x)} = \frac{\theta(A^{l-l_i} x - A^{-l_i} x_i)}{\sum_{j \in J} \theta(A^{l-l_j} x - A^{-l_j} x_j)}.$$

By (5.7) the denominator is $\geq 1$ and has at most $L$ terms by (5.5). Furthermore, for $j \in J$ we have $l - l_j \leq l_i - l_j \leq \omega$ by (5.4). The estimate (5.10) follows now from the iterative application of the quotient rule combined with

(5.11) $$\sup_{x \in \mathbb{R}^n} \sup_{j \leq \omega} \sup_{|\alpha| \leq N} |\partial^\alpha (\theta(A^j \cdot))(x)| = C < \infty.$$

To finish the proof we must show (5.11).

For a given $C^\infty$ function $\Theta$ on $\mathbb{R}^n$ and an integer $k \geq 0$, $\mathfrak{D}^k \Theta(x)$ denotes the derivative of $\Theta$ of order $k$ at the point $x$ thought of as a symmetric, multilinear functional, i.e., $\mathfrak{D}^k \Theta(x) : (\mathbb{R}^n)^k = \mathbb{R}^n \times \ldots \times \mathbb{R}^n \to \mathbb{R}$. The norm is given by

$$||\mathfrak{D}^k \Theta(x)|| = \sup_{\substack{v_i \in \mathbb{R}^n, |v_i| = 1 \\ i = 1, \ldots, k}} |\mathfrak{D}^k \Theta(x)(v_1, \ldots, v_k)|.$$

Suppose $\{e_1, \ldots, e_n\}$ is the standard basis of $\mathbb{R}^n$, and $\sigma$ is any mapping from $\{1, \ldots, k\}$ to $\{1, \ldots, n\}$. Then

$$\Theta(x)(e_{\sigma(1)}, \ldots, e_{\sigma(k)}) = \frac{\partial^{\alpha_1}}{\partial x_1^{\alpha_1}} \cdots \frac{\partial^{\alpha_n}}{\partial x_n^{\alpha_n}} \Theta(x) = \partial^{(\alpha_1, \ldots, \alpha_n)} \Theta(x),$$

where $\alpha_j = \#\{i : \sigma(i) = j\}$. If $\Theta(x) = \theta(A^j x)$ then by the chain rule for any integer $k \geq 1$

$$\mathfrak{D}^k \Theta(x)(v_1, \ldots, v_k) = \mathfrak{D}^k \theta(A^j x)(A^j v_1, \ldots, A^j v_k),$$

for any vectors $v_1, \ldots, v_k \in \mathbb{R}^n$.

For an integer $j \leq \omega$ denote $\Theta(x) = \theta(A^j x)$. Then

$$\sup_{x \in \mathbb{R}^n} \sup_{|\alpha| \leq N} |\partial^\alpha \Theta(x)| \leq \sup_{x \in \mathbb{R}^n} \sup_{|\alpha| \leq N} ||\mathfrak{D}^{|\alpha|} \Theta(x)|| \leq \sup_{x \in \mathbb{R}^n} \sup_{|\alpha| \leq N} ||\mathfrak{D}^{|\alpha|} \theta(A^j x)|| \, ||A^j||^{|\alpha|}$$

$$\leq (\sup_{j \leq \omega} ||A^j||)^N \sup_{x \in \mathbb{R}^n} \sup_{k = 0, \ldots, N} ||\mathfrak{D}^k \theta(x)|| = C < \infty.$$

Indeed, the above suprema are finite since for $j \leq \omega$ we have $||A^j|| \leq c(\lambda_+)^\omega$ by (2.1) and (2.2), and $\theta$ is $C^\infty$ with compact support. Thus, (5.11) holds and hence (5.10). $\square$

LEMMA 5.3. *There exists a constant $A_2 > 0$ independent of $f \in \mathcal{S}'$, $i \in \mathbb{N}$, and $\lambda > 0$ so that*

(5.12) $$\sup_{y \in \mathbb{R}^n} |P_i(y) \zeta_i(y)| \leq A_2 \lambda.$$

PROOF. Let $\pi_1, \ldots, \pi_m$ ($m = \dim \mathcal{P}_s$) be an orthonormal basis of $\mathcal{P}_s$ with respect to the norm (5.9). We have

(5.13) $$P_i = \sum_{k=1}^m \left( \frac{1}{\int \zeta_i} \int f(x) \pi_k(x) \zeta_i(x) dx \right) \overline{\pi_k},$$

where the integral is understood as $\langle f, \pi_k \zeta_i \rangle$. Hence

(5.14) $$1 = \frac{1}{\int \zeta_i} \int |\pi_k(x)|^2 \zeta_i(x) dx \geq \frac{1}{|B_{l_i + \omega}|} \int_{x_i + B_{l_i - \omega}} |\pi_k(x)|^2 \zeta_i(x) dx$$

$$= b^{-\omega} \int_{B_{-\omega}} |\tilde{\pi}_k(x)|^2 dx,$$

where $\tilde{\pi}_k(x) = \pi_k(x_i + A^{l_i}x)$. Since $\mathcal{P}_s$ is finite dimensional all norms on $\mathcal{P}_s$ are equivalent, there exists $C_1 > 0$ such that

$$\sup_{|\alpha| \le s} \sup_{z \in B_\omega} |\partial^\alpha P(z)| \le C_1 \left( \int_{B_{-\omega}} |P(z)|^2 dz \right)^{1/2} \quad \text{for all } P \in \mathcal{P}_s.$$

Therefore by (5.14),

(5.15) $$\sup_{|\alpha| \le s} \sup_{z \in B_\omega} |\partial^\alpha \tilde{\pi}_k(z)| \le C_1 b^{\omega/2} \quad \text{for } k = 1, \ldots, m.$$

For $k = 1, \ldots, m$ define

$$\Phi_k(y) = \frac{b^{l_i}}{\int \zeta_i} \pi_k(z - A^{l_i}y) \zeta_i(z - A^{l_i}y),$$

where $z$ is some point in $(x_i + B_{l_i+\gamma}) \cap \Omega^c$ by (5.3).

We claim there exists a constant $C_2 > 0$ such that $1/C_2 \Phi_k \in \mathcal{S}_N$ for all $k = 1, \ldots, m$. Indeed, if $y \in \operatorname{supp} \Phi_j$, then $z - A^{l_i}y \in x_i + B_{l_i+\omega}$, $-A^{l_i}y \in x_i - z + B_{l_i+\omega} \subset -B_{l_i+\gamma} + B_{l_i+\omega}$, so $y \in B_\gamma - B_\omega$. Hence $\operatorname{supp} \Phi_k$ is bounded.

We can write

$$\Phi_k(y) = \frac{b^{l_i}}{\int \zeta_i} \pi_k(x_i + \tilde{z} - A^{l_i}y) \zeta_i(x_i + \tilde{z} - A^{l_i}y) = \frac{b^{l_i}}{\int \zeta_i} \tilde{\pi}_k(A^{-l_i}\tilde{z} - y) \tilde{\zeta}_i(A^{-l_i}\tilde{z} - y),$$

where $\tilde{\pi}_k(x) = \pi_k(x_i + A^{l_i}x)$, $\tilde{\zeta}_i(x) = \zeta_i(x_i + A^{l_i}x)$ and $z = x_i + \tilde{z}$. Consider

$$\tilde{\zeta}_i(x) = \zeta_i(x_i + A^{l_i}x) = \frac{\theta_i(x_i + A^{l_i}x)}{\sum_j \theta_j(x_i + A^{l_i}x)} = \frac{\theta(x)}{\sum_j \theta(A^{-l_j}(x_i - x_j) + A^{l_i - l_j}x)}.$$

Clearly $\operatorname{supp} \tilde{\zeta}_i \subset B_\omega$ and by Lemma 5.2

$$\sup_{|\alpha| \le N} \sup_{z \in \mathbb{R}^n} |\partial^\alpha \tilde{\zeta}_i(z)| \le A_1.$$

By the product rule, (5.15) and $\operatorname{supp} \tilde{\zeta}_i \subset B_\omega$ we can find a constant $C_3$ so that

$$\sup_{|\alpha| \le N} \sup_{z \in \mathbb{R}^n} |\partial^\alpha (\tilde{\pi}_k(z) \tilde{\zeta}_i(z))| \le C_3.$$

Since $b^{l_i}/\int \zeta_i \le b^{l_i}/b^{l_i-\omega} = b^\omega$, $\operatorname{supp} \Phi_k \subset B_\gamma + B_\omega$, by the above estimate and the definition of $\Phi_k$ we can find a constant $C_2$ so that $\|\Phi_k\|_{\mathcal{S}_N} \le C_2$.

Since

$$(f * (\Phi_k)_{l_i})(z) = \int f(y) b^{-l_i} \Phi_k(A^{-l_i}(z - y)) dy$$

$$= \frac{1}{\int \zeta_i} \int f(y) \pi_k(z - A^{l_i}(A^{-l_i}(z - y))) \zeta_i(z - A^{l_i}(A^{-l_i}(z - y))) dy$$

$$= \frac{1}{\int \zeta_i} \int f(y) \pi_k(y) \zeta_i(y) dy,$$

we have

$$\left| \frac{1}{\int \zeta_i} \int f(y) \pi_k(y) \zeta_i(y) dy \right| \le Mf(z) \|\Phi_k\|_{\mathcal{S}_N} \le C_2 \lambda.$$

By (5.13), (5.15) and the above estimate

$$
\sup_{z \in x_i + B_{l_i+\omega}} |P_i(z)| \leq mC_2 \lambda C_1 b^{\omega/2}, \tag{5.16}
$$

and therefore we have (5.12) with $A_2 = mb^{\omega/2}C_1C_2$. □

LEMMA 5.4. *There exists a constant $A_3 > 0$ such that*

$$
Mb_i(x) \leq A_3 Mf(x) \quad \text{for } x \in x_i + B_{l_i+2\omega}. \tag{5.17}
$$

PROOF. Take $\varphi \in \mathcal{S}_N$, and $x \in x_i + B_{l_i+2\omega}$.
**Case I.** For $l \leq l_i$ we write

$$
(b_i * \varphi_l)(x) = (f * \Phi_l)(x) - ((P_i \zeta_i) * \varphi_l)(x), \tag{5.18}
$$

where $\Phi(z) := \varphi(z)\zeta_i(x - A^l z)$. Define $\tilde{\zeta}_i(z) := \zeta_i(x - A^l z)$. By Lemma 5.2 we have

$$
\sup_{|\alpha| \leq N} \sup_{z \in \mathbb{R}^n} |\partial^\alpha \tilde{\zeta}_i(z)| \leq A_1.
$$

By the product rule there is a constant $C$ depending only on $N$ so that

$$
\begin{aligned}
\|\Phi\|_{\mathcal{S}_N} &= \sup_{z \in \mathbb{R}^n} \max(1, \rho(z)^N) \sup_{|\alpha| \leq N} |\partial^\alpha \Phi(z)| \\
&= \sup_{z \in \mathbb{R}^n} \max(1, \rho(z)^N) \sup_{|\alpha| \leq N} |\partial^\alpha (\varphi \tilde{\zeta}_i)(z)| \\
&\leq C \sup_{z \in \mathbb{R}^n} \max(1, \rho(z)^N) \sup_{|\alpha| \leq N} |\partial^\alpha \varphi(z)| \sup_{|\alpha| \leq N} |\partial^\alpha \tilde{\zeta}_i(z)| \leq A_1 C.
\end{aligned}
$$

Note that for $N \geq 2$ there is a constant $C' > 0$ so that $\|\varphi\|_1 \leq C'$ for all $\varphi \in \mathcal{S}_N$. Therefore, by Lemma 5.3 and (5.18), we have

$$
\begin{aligned}
|b_i * \varphi_l(x)| &\leq \|\Phi\|_{\mathcal{S}_N} Mf(x) + A_2 \lambda \|\varphi\|_1 \leq A_1 C Mf(x) + A_2 C' \lambda \\
&\leq (A_1 C + A_2 C') Mf(x),
\end{aligned}
$$

since $Mf(x) > \lambda$ for $x \in \Omega$.
**Case II.** For $l > l_i$ by a simple calculation we can write

$$
(b_i * \varphi_l)(x) = b^{l_i - l}(f * \Phi_{l_i})(x) - ((P_i \zeta_i) * \varphi_l)(x),
$$

where $\Phi(z) := \varphi(A^{l_i-l}z)\zeta_i(x - A^{l_i}z)$. Define $\tilde{\varphi}(z) := \varphi(A^{l_i-l}z)$ and $\tilde{\zeta}_i(z) := \zeta_i(x - A^{l_i}z)$. If $z \in \operatorname{supp} \Phi$ then $x - A^{l_i}z \in x_i + B_{l_i+\omega}$, so $A^{l_i}z \in x - x_i + B_{l_i+\omega} \subset B_{l_i+2\omega} + B_{l_i+\omega} \subset B_{l_i+3\omega}$. Hence, $\operatorname{supp} \Phi \subset B_{3\omega}$. By Lemma 5.2 and since $\|A^{l_i-l}\| \leq c$ by (2.2) we can find a constant $C > 0$ independent of $l > l_i$ so that

$$
\sup_{|\alpha| \leq N} \sup_{z \in \mathbb{R}^n} |\partial^\alpha \tilde{\varphi}(z)| \leq C, \quad \sup_{|\alpha| \leq N} \sup_{z \in \mathbb{R}^n} |\partial^\alpha \tilde{\zeta}_i(z)| \leq A_1.
$$

By the product rule and the boundedness of the support of $\Phi$ we can find a constant $C''$ so that $\|\Phi\|_{\mathcal{S}_N} \leq C''$. Let $C'$ be the constant such that $\|\varphi\|_1 \leq C'$ for $\varphi \in \mathcal{S}_N$ for $N \geq 2$. As in the case I

$$
|(b_i * \varphi_l)(x)| \leq b^{l_i-l} \|\Phi\|_{\mathcal{S}_N} Mf(x) + A_2 \lambda \|\varphi\|_1 \leq (C'' + A_2 C') Mf(x).
$$

By combining both cases we arrive at (5.17). □

LEMMA 5.5. *Suppose $Q \subset \mathbb{R}^n$ is bounded, convex, and $0 \in Q$, and $N$ is a positive integer. Then there is a constant $C$ depending only on $Q$ and $N$ such that for every $\phi \in \mathcal{S}$ and every integer $s$, $0 \le s < N$ we have*

$$(5.19) \qquad \sup_{z \in Q} \sup_{|\alpha| \le N} |\partial^\alpha R_y(z)| \le C \sup_{z \in y+Q} \sup_{s+1 \le |\alpha| \le N} |\partial^\alpha \phi(z)|,$$

*where $R_y$ is the remainder of the Taylor expansion of $\phi$ of order $s$ at the point $y \in \mathbb{R}^n$.*

PROOF. We write the Taylor expansion of $\phi$ of order $s$ at the point $y$

$$\phi(y+z) = \sum_{|\beta| \le s} \frac{\partial^\beta \phi(y)}{\beta!} z^\beta + R_y(z).$$

For any multi-index $|\alpha| \le s$ we have

$$\partial^\alpha R_y(z) = \partial^\alpha \phi(y+z) - \sum_{\substack{|\beta| \le s, \\ \beta \ge \alpha}} \frac{\partial^\beta \phi(y)}{(\beta-\alpha)!} z^{\beta-\alpha} = \partial^\alpha \phi(y+z) - \sum_{|\beta| \le s-|\alpha|} \frac{\partial^{\alpha+\beta}\phi(y)}{\beta!} z^\beta,$$

where $(\alpha_1, \ldots, \alpha_n) = \alpha \le \beta = (\beta_1, \ldots, \beta_n)$ means $\alpha_i \le \beta_i$ for all $i = 1, \ldots, n$. We used here

$$\frac{1}{\beta!} \partial^\beta z^\alpha = \begin{cases} z^{\beta-\alpha}/(\beta-\alpha)! & \alpha \le \beta, \\ 0 & \text{otherwise.} \end{cases}$$

Therefore, $\partial^\alpha R_y$ is the Taylor remainder of $\partial^\alpha \phi$ of order $s - |\alpha|$ expanded at $y$ and by the Taylor Remainder Theorem

$$|\partial^\alpha R_y(z)| \le C \sup_{w \in [y, y+z]} \sup_{|\beta|=s-|\alpha|+1} |\partial^{\alpha+\beta}\phi(w)||z|^{s-|\alpha|+1},$$

where $[y, y+z]$ is the line segment connecting $y$ and $y+z$. Thus,

$$\sup_{z \in Q} |\partial^\alpha R_y(z)| \le C' \sup_{w \in y+Q} \sup_{|\alpha|=s+1} |\partial^\alpha \phi(w)|.$$

If $s < |\alpha| \le N$ then $\partial^\alpha R_y(z) = \partial^\alpha \phi(y+z)$, and the estimate (5.19) follows immediately. $\square$

LEMMA 5.6. *Suppose $0 \le s < N$. Then there exists a constant $A_4 > 0$ so that for $i \in \mathbb{N}$ and all integers $t \ge 0$*

$$(5.20) \qquad Mb_i(x) \le A_4 \lambda \lambda_-^{-t(s+1)} \qquad \text{for } x \in x_i + B_{t+l_i+2\omega+1} \setminus B_{t+l_i+2\omega}.$$

PROOF. Suppose $\varphi \in \mathcal{S}_N$, and $l \in \mathbb{Z}$. Pick some $w \in (x_i + B_{l_i+\gamma}) \cap \Omega^c$.
**Case I.** For $l \le l_i$ we have

$$(5.21) \qquad b_i * \varphi_l(x) = f * \Phi_l(w) - (P_i \zeta_i) * \varphi_l(x),$$

where $\Phi(z) := \varphi(z + A^{-l}(x-w))\zeta_i(w - A^l z)$. If $z \in \operatorname{supp} \Phi$ then $z \in \operatorname{supp} \zeta_i(w - A^l \cdot)$, i.e., $w - A^l z \in x_i + A^{l_i} B_\omega$, so $z \in A^{l_i-l}(B_\omega + A^{-l_i}(w-x_i)) \subset A^{l_i-l}(B_\omega - B_\gamma) \subset$

## 5. THE CALDERÓN-ZYGMUND DECOMPOSITION

$A^{l_i-l}B_{\gamma+\omega}$. Suppose $x \in x_i + B_{t+l_i+2\omega+1} \setminus B_{t+l_i+2\omega}$ for some integer $t \geq 0$. If $z \in \operatorname{supp} \Phi$ then $z \in A^{l_i-l}B_\omega + A^{-l}(w - x_i)$ and by (2.11)

$$
\begin{aligned}
z + A^{-l}(x - w) &\in A^{l_i-l}B_\omega + A^{-l}(w - x_i) + A^{-l}(x - w) \\
&= A^{l_i-l}B_\omega + A^{-l}(x - x_i) \subset A^{l_i-l}(B_\omega + B_{t+2\omega+1} \setminus B_{t+2\omega}) \\
&\subset A^{l_i-l}((B_{t+2\omega})^{\mathbf{c}} + B_\omega) \subset A^{l_i-l}((B_{t+\omega})^{\mathbf{c}}) = (B_{t+l_i-l+\omega})^{\mathbf{c}}.
\end{aligned}
\tag{5.22}
$$

By Lemma 5.2 the partial derivatives of $\zeta_i(w - A^l \cdot)$ of order up to $N$ are bounded by $A_1$. Using $\operatorname{supp} \Phi \subset B_{l_i-l+\gamma+\omega}$, $\|\varphi\|_{\mathcal{S}_N} \leq 1$, (5.22) and the product rule we have

$$
\begin{aligned}
\|\Phi\|_{\mathcal{S}_N} &= \sup_{z \in \operatorname{supp} \Phi} \sup_{|\alpha| \leq N} \max(1, \rho(z)^N) |\partial^\alpha \Phi(z)| \\
&\leq (b^{l_i-l+\gamma+\omega})^N A_1 C \sup_{|\alpha| \leq N} \sup_{z \in \operatorname{supp} \Phi} |\partial^\alpha \varphi(z + A^{-l}(x - w))| \\
&\leq A_1 C b^{N(l_i-l+\gamma+\omega)} \sup_{z \in A^{l_i-l}B_\omega + A^{-l}(w-x_i)} \max(1, \rho(z + A^{-l}(x - w)))^{-N} \\
&\leq A_1 C b^{N(l_i-l+\gamma+\omega)} b^{-N(t+l_i-l+\omega)} = A_1 C b^{N\gamma} b^{-Nt}.
\end{aligned}
\tag{5.23}
$$

This enables us to estimate the first term in (5.21). For the second term, note that if $x \in x_i + (B_{t+l_i+2\omega})^{\mathbf{c}}$ for some integer $t \geq 0$, and $y \in x_i + B_{l_i+\omega}$ then $A^{-l}(x - y) \in A^{l_i-l}((B_{t+2\omega})^{\mathbf{c}} - B_\omega) \subset (B_{t+l_i-l+\omega})^{\mathbf{c}}$ by (2.11). Since $\|\varphi\|_{\mathcal{S}_N} \leq 1$

$$
\begin{aligned}
|(P_i \zeta_i) * \varphi_l(x)| &= \left| \int (P_i \zeta_i)(y) b^{-l} \varphi(A^{-l}(x - y)) dy \right| \\
&\leq b^{-l} \int_{x_i + B_{l_i+\omega}} |(P_i \zeta_i)(y)| |\varphi(A^{-l}(x - y))| dy \\
&\leq b^{-l} b^{l_i+\omega} A_2 \lambda (b^{t+l_i-l+\omega})^{-N} \leq A_2 \lambda b^{-Nt}.
\end{aligned}
$$

Combining the above, (5.21) and (5.23) we obtain

$$
|b_i * \varphi_l(x)| \leq \|\Phi\|_{\mathcal{S}_N} Mf(w) + A_2 \lambda b^{-Nt} \leq (A_1 C b^{N\gamma} + A_2) \lambda b^{-Nt}.
$$

**Case II.** For fixed $l > l_i$ and $\varphi \in \mathcal{S}_N$ define $\phi(z) = \varphi(A^{l_i-l}z)$. Suppose $x \in x_i + B_{t+l_i+2\omega+1} \setminus B_{t+l_i+2\omega}$ for some integer $t \geq 0$. Consider the Taylor expansion of $\phi \in \mathcal{S}$ of order $s$ at the point $y := A^{-l_i}(x - w)$,

$$
\phi(y + z) = \sum_{|\alpha| \leq s} \frac{\partial^\alpha \phi(y)}{\alpha!} z^\alpha + R_y(z),
$$

where $R_y$ denotes the reminder. Since the distribution $b_i$ annihilates polynomials of degree $\leq s$ we have

$$
\begin{aligned}
(b_i * \varphi_l)(x) &= b^{-l} \int b_i(z) \varphi(A^{-l}(x - z)) dz = b^{-l} \int b_i(z) \phi(A^{-l_i}(x - z)) dz \\
&= b^{-l} \int b_i(z) R_{A^{-l_i}(x-w)}(A^{-l_i}(w - z)) dz \\
&= b^{l_i-l}(f * \Phi_{l_i})(w) + b^{-l} \int P_i(z) \zeta_i(z) R_{A^{-l_i}(x-w)}(A^{-l_i}(w - z)) dz,
\end{aligned}
\tag{5.24}
$$

where

$$
\Phi(z) := R_{A^{-l_i}(x-w)}(z) \zeta_i(w - A^{l_i} z).
\tag{5.25}
$$

If $z \in \operatorname{supp} \Phi$ then $w - A^{l_i}z \in x_i + B_{l_i+\omega}$, so $z \in B_\gamma + B_\omega$ and $\operatorname{supp} \Phi \subset B_{\gamma+\omega}$.

Apply Lemma 5.5 to $\phi(z) = \varphi(A^{l_i-l}z)$, $y = A^{-l_i}(x-w)$ and $Q = B_{\gamma+\omega}$. First assume $t \geq \gamma + \omega$ and $x \in x_i + B_{t+l_i+2\omega+1} \setminus B_{t+l_i+2\omega}$. Since

$$y + B_{\gamma+\omega} \subset A^{-l_i}((B_{t+l_i+2\omega+1} \setminus B_{t+l_i+2\omega}) + B_{l_i+\gamma}) + B_{\gamma+\omega}$$
$$= (B_{t+2\omega+1} \setminus B_{t+2\omega}) + B_{\gamma+2\omega} \subset (B_{t+2\omega})^{\mathbf{c}} + B_{t+\omega} \subset (B_{t+\omega})^{\mathbf{c}},$$

we have

$$\sup_{z \in B_{\gamma+\omega}} \sup_{|\alpha| \leq N} |\partial^\alpha R_y(z)| \leq C \sup_{z \in y+B_{\gamma+\omega}} \sup_{s+1 \leq |\alpha| \leq N} |\partial^\alpha \phi(z)|$$
$$\leq C \sup_{z \in y+B_{\gamma+\omega}} \sup_{s+1 \leq |\alpha| \leq N} \lambda_-^{(l_i-l)(|\alpha|+1)} |\partial^\alpha \varphi(A^{l_i-l}z)|$$
$$\leq C\lambda_-^{(l_i-l)(s+1)} \sup_{z \in y+B_{\gamma+\omega}} \sup_{s+1 \leq |\alpha| \leq N} |\partial^\alpha \varphi(A^{l_i-l}z)|$$
$$\leq C\lambda_-^{(l_i-l)(s+1)} \sup_{z \in (B_{t+\omega})^{\mathbf{c}}} \max(1, \rho(A^{l_i-l}z))^{-N}$$
$$= C\lambda_-^{(l_i-l)(s+1)} \min(1, b^{-N(l_i-l+t+\omega)}).$$

Since $\lambda_-^{s+1} b^{-N} \leq 1$ the last expression is maximized over $l > l_i$ if $l_i - l + t + \omega = 0$, i.e., $l = l_i + t + \omega$. Therefore, we have for $t \geq \gamma + \omega$

$$\sup_{z \in B_{\gamma+\omega}} \sup_{|\alpha| \leq N} |\partial^\alpha R_y(z)| \leq C\lambda_-^{-t(s+1)}.$$

If $0 \leq t < \gamma + \omega$ and $x \in B_{t+l_i+2\omega+1} \setminus B_{t+l_i+2\omega}$ we can estimate as before

$$\sup_{z \in B_{\gamma+\omega}} \sup_{|\alpha| \leq N} |\partial^\alpha R_y(z)| \leq C\lambda_-^{(l_i-l)(s+1)} \sup_{z \in y+B_{\gamma+\omega}} \sup_{s+1 \leq |\alpha| \leq N} |\partial^\alpha \varphi(A^{l_i-l}z)| \leq C.$$

Therefore, we have for all $t \geq 0$

$$(5.26) \qquad \sup_{z \in B_{\gamma+\omega}} \sup_{|\alpha| \leq N} |\partial^\alpha R_y(z)| \leq C\lambda_-^{(\gamma+\omega)(s+1)} \lambda_-^{-t(s+1)}.$$

By (5.25), (5.26), Lemma 5.2 and the product rule we can find a constant $C'$ so that

$$\|\Phi\|_{\mathcal{S}_N} \leq C'\lambda_-^{-t(s+1)} \qquad \text{for } t \geq 0.$$

Therefore by (5.24),

$$|(b_i * \varphi_l)(x)| \leq b^{l_i - l}|f * \Phi_{l_i}(w)| + b^{-l} \int |P_i(z)\zeta_i(z) R_{A^{-l_i}(x-w)}(A^{-l_i}(w-z))|dz$$
$$\leq b^{l_i-l} Mf(w) \|\Phi\|_{\mathcal{S}_N} + b^{-l} b^{l_i+\omega} A_2 \lambda \sup_{z \in B_{\gamma+\omega}} |R_{A^{-l_i}(x-w)}(z)|.$$

Thus

$$|(b_i * \varphi_l)(x)| \leq C\lambda \lambda_-^{-t(s+1)} \qquad \text{for } x \in B_{t+l_i+2\omega+1} \setminus B_{t+l_i+\omega},\ t \geq 0.$$

This ends the proof of case II. Combining the two cases we arrive at the estimate

$$|(b_i * \varphi_l)(x)| \leq A_4 \lambda \lambda_-^{-t(s+1)} \qquad \text{for all } l \in \mathbb{Z}.$$

Hence we have (5.20). $\square$

The above proof also works in the case $s \geq N$ and yields the estimate

$$Mb_i(x) \leq A_4 \lambda \lambda_-^{-tN} \qquad \text{for } x \in x_i + B_{t+l_i+2\omega+1} \setminus B_{t+l_i+2\omega},$$

for all $i \in \mathbb{N}$ and all integers $t \geq 0$.

LEMMA 5.7. *Suppose $0 < p \leq 1$, $s \geq \lfloor \ln b/(p \ln \lambda_-) \rfloor$, $N > s$ and $f \in H^p$, where $H^p$ is given by (3.23). Then there exists a constant $A_5$ independent of $f \in H^p$, $i \in \mathbb{N}$, and $\lambda > 0$ such that*

$$(5.27) \qquad \int_{\mathbb{R}^n} Mb_i(x)^p dx \leq A_5 \int_{x_i + B_{l_i + 2\omega}} Mf(x)^p dx.$$

*Moreover, the series $\sum_i b_i$ converges in $H^p$, and*

$$(5.28) \qquad \int_{\mathbb{R}^n} M\left(\sum_i b_i\right)(x)^p dx \leq LA_5 \int_\Omega Mf(x)^p dx,$$

*where $L$ is as in (5.5).*

PROOF. By Lemma 5.4

$$\int_{\mathbb{R}^n} Mb_i(x)^p dx \leq (A_3)^p \int_{x_i + B_{l_i+2\omega}} Mf(x)^p dx + \int_{(x_i + B_{l_i+2\omega})^c} Mb_i(x)^p dx.$$

By Lemma 5.6

$$\int_{(x_i+B_{l_i+2\omega})^c} Mb_i(x)^p dx = \sum_{t=0}^\infty \int_{x_i+B_{t+l_i+2\omega+1} \setminus B_{t+l_i+2\omega}} Mb_i(x)^p dx$$

$$\leq \sum_{t=0}^\infty b^{t+l_i+2\omega+1}(A_4)^p \lambda^p \lambda_-^{-t(s+1)p} \leq (A_4)^p b \sum_{t=0}^\infty b^t \lambda_-^{-t(s+1)p} \int_{x_i+B_{l_i+2\omega}} Mf(x)^p dx$$

$$\leq C' \int_{x_i+B_{l_i+2\omega}} Mf(x)^p dx,$$

because $Mf(x) > \lambda$ for $x \in x_i + B_{l_i+2\omega}$ and $\lambda_-^{(s+1)p} > b$. Combining the last two estimates we obtain (5.27).

Since $H^p$ is complete and

$$\sum_i \int_{\mathbb{R}^n} Mb_i(x)^p dx \leq A_5 \sum_i \int_{x_i+B_{l_i+2\omega}} Mf(x)^p dx \leq LA_5 \int_\Omega Mf(x)^p dx,$$

$\sum_i b_i$ converges in $H^p$. Therefore, $\sum_i b_i$ converges in $\mathcal{S}'$ and as in the proof of Proposition 3.12 we obtain $M(\sum_i b_i)(x) \leq \sum_i Mb_i(x)$ and thus (5.28) holds. $\square$

LEMMA 5.8. *Suppose $s \geq 0$, $N \geq 2$, and $f \in L^1(\mathbb{R}^n)$. Then the series $\sum_{i \in \mathbb{N}} b_i$ converges in $L^1(\mathbb{R}^n)$. Moreover, there is a constant $A_6$, independent of $f$, $i$, $\lambda$, such that*

$$(5.29) \qquad \int_{\mathbb{R}^n} \sum_{i \in \mathbb{N}} |b_i(x)| dx \leq A_6 \int_{\mathbb{R}^n} |f(x)| dx.$$

PROOF. By Lemma 5.3

$$\int_{\mathbb{R}^n} |b_i(x)|dx = \int_{\mathbb{R}^n} |(f(x) - P_i(x))\zeta_i(x)|dx$$
$$\leq \int_{x_i+B_{l_i+\omega}} |f(x)|dx + \int_{x_i+B_{l_i+\omega}} |P_i(x)\zeta_i(x)|dx \leq \int_{x_i+B_{l_i+\omega}} |f(x)|dx + A_2\lambda b^{l_i+\omega}.$$

Therefore, by (5.2) and (5.5)

$$\sum_{i\in\mathbb{N}} \int_{\mathbb{R}^n} |b_i(x)|dx \leq L \int_\Omega |f(x)|dx + A_2 b^{2\omega}\lambda|\Omega| \leq A_6 \int_{\mathbb{R}^n} |f(x)|dx,$$

by the Maximal Theorem 3.6. □

LEMMA 5.9. *Suppose $0 \leq s < N$ and $\sum_{i\in\mathbb{N}} b_i$ converges in $\mathcal{S}'$. There is a constant $A_7$, independent of $f \in \mathcal{S}'$, $\lambda > 0$, such that*

$$(5.30) \qquad Mg(x) \leq A_7\lambda \sum_i \lambda_-^{-t_i(s+1)} + Mf(x)\mathbf{1}_{\Omega^c}(x),$$

*where*

$$(5.31) \qquad t_i = t_i(x) := \begin{cases} t & \text{if } x \in B_{t+l_i+2\omega+1} \setminus B_{t+l_i+2\omega} \text{ for some } t \geq 0, \\ 0 & \text{otherwise.} \end{cases}$$

PROOF. If $x \notin \Omega$ then as in the proof of Proposition 3.12 we have

$$Mg(x) \leq Mf(x) + \sum_{i\in\mathbb{N}} Mb_i(x) \leq Mf(x) + \sum_{i\in\mathbb{N}} A_4\lambda\lambda_-^{-t_i(s+1)},$$

by Lemma 5.6—which is exactly what we need.

If $x \in \Omega$ choose $k \in \mathbb{N}$ such that $x \in x_k + B_{l_k}$. Let $J := \{i \in \mathbb{N} : (x_i + B_{l_i+2\omega}) \cap (x_k + B_{l_k+2\omega}) \neq \emptyset\}$. By (5.5) the cardinality of $J$ is bounded by $L$. By Lemma 5.6 we have

$$(5.32) \qquad \sum_{i\notin J} Mb_i(x) \leq A_4\lambda \sum_{i\notin J} \lambda_-^{-t_i(s+1)}.$$

It suffices to estimate the grand maximal function of $g + \sum_{i\notin J} b_i = f - \sum_{i\in J} b_i$.

Take $\varphi \in \mathcal{S}_N$ and $l \in \mathbb{Z}$.

**Case I.** For $l \leq l_k$ we write

$$(5.33) \qquad \begin{aligned}(f - \sum_{i\in J} b_i) * \varphi_l(x) &= (f\eta) * \varphi_l(x) + (\sum_{i\in J} P_i\zeta_i) * \varphi_l(x) \\ &= f * \Phi_l(w) + (\sum_{j\in J} P_i\zeta_i) * \varphi_l(x),\end{aligned}$$

where $w \in (x_k + B_{l_k+\gamma}) \cap \Omega^c$, $\eta = 1 - \sum_{i\in J} \zeta_i$, and

$$(5.34) \qquad \Phi(z) := \varphi(z + A^{-l}(x-w))\eta(w - A^l z).$$

By the definition of $J$, $\eta \equiv 0$ on $x_k + B_{l_k+2\omega}$. Thus, if $z \in \operatorname{supp} \Phi$ then $w - A^l z \in (x_k + B_{l_k+2\omega})^c$, and $z + A^{-l}(x-w) = A^{-l}(A^l z - w + x) \subset A^{-l}(-x_k + $

$(B_{l_k+2\omega})^c + x_k + B_{l_k}) \subset A^{-l}(B_{l_k+\omega})^c = (B_{l_k-l+\omega})^c$ by (2.11). Since $A^{-l}(x-w) \subset A^{-l}(B_{l_k} - B_{l_k+\gamma}) = B_{l_k-l} - B_{l_k-l+\gamma} \subset B_{l_k-l+\gamma+\omega}$ we have

(5.35) $\quad \rho(z) \leq b^\omega(\rho(z + A^{-l}(x-w)) + \rho(-A^{-l}(x-w))) \leq c'\rho(z + A^{-l}(x-w)),$

for $z \in \operatorname{supp} \Phi$, where $c' = 2b^{\gamma+\omega}$. By (5.4) $l_i \geq l_k - \omega$ hence $l \leq l_i + \omega$. By Lemma 5.2 applied for every $i \in J$, the chain rule, and (5.5) the derivatives of $\eta(w - A^l \cdot)$ of order $\leq s$ are bounded by some universal constant $C$. For $z \in \operatorname{supp} \Phi$ by the product rule, (5.34), and (5.35)

$$\max(1, \rho(z)^N) \sup_{|\alpha| \leq N} |\partial^\alpha \Phi(z)| \leq C \max(1, \rho(z)^N) \sup_{|\alpha| \leq N} |\partial^\alpha \varphi(z + A^{-l}(x-w))|$$
$$\leq C \max(1, \rho(z)^N) / \max(1, \rho(z + A^{-l}(x-w))^N) \leq C(c')^N,$$

and therefore $\|\Phi\|_{\mathcal{S}_N} \leq C(c')^N$. Hence

$$|(f * \Phi_l)(w)| \leq Mf(w)\|\Phi\|_{\mathcal{S}_N} \leq C(c')^N \lambda.$$

Since for $N \geq 2$ there is a constant $C' > 0$ so that $\|\varphi\|_1 \leq C'$ for all $\varphi \in \mathcal{S}_N$ and Lemma 5.2

$$\left|\left(\sum_{i \in J} P_i \zeta_i\right) * \varphi_l(x)\right| \leq LA_2 C' \lambda.$$

By the two estimates above and (5.33) we have $|(f - \sum_{i \in J} b_i) * \varphi_l(x)| \leq (C(c')^N + LA_2 C')\lambda$.

**Case II.** For $l > l_k$ we let $\Phi(z) := \varphi(z + A^{-l}(x-w))$, where $w \in (x_k + B_{l_k+\gamma}) \cap \Omega^c$. Since $A^{-l}(x-w) \in B_{l_k-l} - B_{l_k-l+\gamma} \subset B_0 - B_\gamma \subset B_{\gamma+\omega}$ we have $\|\Phi\|_{\mathcal{S}_N} \leq b^{\gamma+2\omega}$. Therefore,

$$|(f * \varphi_l)(x)| = |(f * \Phi_l)(w)| \leq Mf(w)\|\Phi\|_{\mathcal{S}_N} \leq b^{\gamma+2\omega}\lambda.$$

By Lemma 5.6

$$\sum_{i \in J} |(b_i * \varphi_l)(x)| = \sum_{i \in J} |(b_i * \Phi_l)(w)| \leq \|\Phi\|_{\mathcal{S}_N} \sum_{i \in J} Mb_i(w) \leq b^{\gamma+2\omega} LA_4 \lambda,$$

because $w \notin x_i + B_{l_i+2\omega}$ for all $i \in \mathbb{N}$. By the two estimates above we have $|(f - \sum_{i \in J} b_i) * \varphi_l(x)| \leq b^{\gamma+2\omega}(LA_4 + 1)\lambda$.

Combining cases I and II we have $M(f - \sum_{i \in J} b_i)(x) \leq C\lambda$, where $C$ is the maximum of the two constants in these cases. Since $t_k = 0$ the $k$th term in the sum (5.30) takes care of the estimate of the maximal function of $f - \sum_{i \in J} b_i$ and by (5.32) we obtain (5.30). $\square$

LEMMA 5.10. *(i) Suppose $0 < p \leq 1$, $s \geq \lfloor \ln b/(p \ln \lambda_-) \rfloor$, $N > s$, and $Mf \in L^p$. Then $Mg \in L^1$, and there exists a constant $A_8$ (independent of $f$ and $\lambda$) such that*

(5.36) $$\int_{\mathbb{R}^n} Mg(x)dx \leq A_8 \lambda^{1-p} \int_{\mathbb{R}^n} Mf(x)^p dx.$$

*(ii) Without an assumption on $N$ ($N \geq 2$), if $f \in L^1$ then $g \in L^\infty$ and there exists a constant $A_9$ (independent of $f$ and $\lambda$) such that $\|g\|_\infty \leq A_9 \lambda$.*

PROOF. (i) By Lemmas 5.7 and 5.9

$$\int_{\mathbb{R}^n} Mg(x)dx \leq A_7\lambda \sum_{i\in\mathbb{N}} \int_{\mathbb{R}^n} \lambda_-^{-t_i(x)(s+1)} dx + \int_{\Omega^c} Mf(x)dx,$$

where $t_i(x)$ is defined as in Lemma 5.9. For fixed $i \in \mathbb{N}$

$$\int_{\mathbb{R}^n} \lambda_-^{-t_i(x)(s+1)} dx = \int_{x_i+B_{l_i+2\omega}} dx + \sum_{t=0}^{\infty} \int_{x_i+B_{t+l_i+2\omega+1}\setminus B_{t+l_i+2\omega}} \lambda_-^{-t(s+1)} dx$$

$$\leq b^{l_i+2\omega} + \sum_{t=0}^{\infty} b^{t+l_i+2\omega+1}\lambda_-^{-t(s+1)} = b^{l_i+2\omega}\left[1 + b\sum_{t=0}^{\infty}(b\lambda_-^{-s-1})^t\right] = C|B_{l_i+2\omega}|,$$

where $C$ represents the value of the expression in the bracket. Taking the sum over all $i \in \mathbb{N}$ we obtain

$$\int_{\mathbb{R}^n} Mg(x)dx \leq A_7\lambda C b^{3\omega} \sum_{i\in\mathbb{N}} |B_{l_i-\omega}| + \int_{\Omega^c} Mf(x)dx$$

$$\leq A_7\lambda C b^{3\omega}|\Omega| + \int_{\Omega^c} Mf(x)dx$$

$$\leq A_7\lambda C b^{3\omega}\lambda^{-p}\int_{\Omega} Mf(x)^p dx + \lambda^{1-p}\int_{\Omega^c} Mf(x)^p dx$$

$$\leq A_8\lambda^{1-p}\int_{\mathbb{R}^n} Mf(x)^p dx.$$

(ii) If $f \in L^1$ then $g$ and $b_i$'s are functions and $\sum_i b_i$ converges in $L^1$ by Lemma 5.8 (and thus in $\mathcal{S}'$). Since

$$g = f - \sum_{i\in\mathbb{N}} b_i = f(1 - \sum_{i\in\mathbb{N}} \zeta_i) + \sum_{i\in\mathbb{N}} P_i\zeta_i = f\mathbf{1}_{\Omega^c} + \sum_{i\in\mathbb{N}} P_i\zeta_i,$$

by Lemma 5.3 and (5.5) for $x \in \Omega$ we have $|g(x)| \leq LA_2\lambda$. Since $|g(x)| = |f(x)| \leq Mf(x) \leq \lambda$ for a.e. $x \in \Omega^c$, therefore $||g||_\infty \leq LA_2\lambda = A_9\lambda$. □

COROLLARY 5.11. *If $0 < p < 1$, $H^p \cap L^1$ is dense in $H^p$, where $H^p$ is given by (3.23) for $N > \lfloor \ln b/(p\ln\lambda_-)\rfloor$.*

PROOF. If $f \in H^p$ and $\lambda > 0$, let $f = g^\lambda + \sum_i b_i^\lambda$ be the Calderón-Zygmund decomposition of $f$ of degree $s$, $\lfloor \ln b/(p\ln\lambda_-)\rfloor \leq s < N$, and height $\lambda$ associated to $Mf = M_N f$. By Lemma 5.7,

$$\left\|\sum_{i\in\mathbb{N}} b_i^\lambda\right\|_{H^p} \leq LA_5\int_{\{x:Mf(x)>\lambda\}} Mf(x)^p dx.$$

Therefore, $g^\lambda \to f$ in $H^p$ as $\lambda \to \infty$. But by Lemma 5.10, $Mg^\lambda \in L^1$, so by Theorem 3.9, $g^\lambda \in L^1$. □

## 6. The atomic decomposition of $H^p$

We will continue to closely follow the proof of atomic decomposition as presented by Folland and Stein in [FoS]. The proofs of Lemmas 6.1 and 6.2 still require some technical arguments, whereas the proofs of Lemma 6.3 and Theorems 6.4 and 6.5 are copied almost verbatim from [FoS], see also [LU], since all the necessary preparatory work is already done.

Suppose $f$ is a tempered distribution such that $Mf = M_N f \in L^p$ for some $0 < p \leq 1$ and $N > s := \lfloor \ln b / (p \ln \lambda_-) \rfloor$. Thus $N \geq N_p$. For each $k \in \mathbb{Z}$ consider the Calderón-Zygmund decomposition of $f$ of degree $s$ and height $\lambda = 2^k$ associated to $Mf$,

$$f = g^k + \sum_{i \in \mathbb{N}} b_i^k, \quad \text{where}$$

(6.1) $\qquad \Omega^k := \{x : Mf(x) > 2^k\}, \quad b_i^k := (f - P_i^k)\zeta_i^k, \quad B_i^k := x_i^k + B_{l_i^k}.$

Recall that for fixed $k \in \mathbb{Z}$, $(x_i = x_i^k)_{i \in \mathbb{N}}$ is a sequence in $\Omega^k$ and $(l_i = l_i^k)_{i \in \mathbb{N}}$ a sequence of integers such that (5.1)–(5.5) hold for $\Omega = \Omega^k$, $\zeta_i = \zeta_i^k$ is given by (5.8) and $P_i = P_i^k$ is the projection of $f$ onto $\mathcal{P}_s$ with respect to the norm given by (5.9).

Define a polynomial $P_{ij}^{k+1}$ as an orthogonal projection of $(f - P_j^{k+1})\zeta_i^k$ on $\mathcal{P}_s$ with respect to the norm

(6.2) $\qquad \|P\|^2 = \dfrac{1}{\int \zeta_j^{k+1}} \int |P(x)|^2 \zeta_j^{k+1}(x) dx,$

that is $P_{ij}^{k+1}$ is the unique element of $\mathcal{P}_s$ such that

$$\int_{\mathbb{R}^n} (f(x) - P_j^{k+1}(x))\zeta_i^k(x) Q(x) \zeta_j^{k+1}(x) = \int_{\mathbb{R}^n} P_{ij}^{k+1}(x) Q(x) \zeta_j^{k+1}(x) dx.$$

Here the first integral is understood as $\langle (f - P_j^{k+1})\zeta_i^k, Q\zeta_j^{k+1}\rangle$ in the case that $f$ is not a regular distribution. For convenience we denote $\hat{B}_i^k := x_i^k + B_{l_i^k + \omega}$.

**LEMMA 6.1.** *(i) If $\hat{B}_j^{k+1} \cap \hat{B}_i^k \neq \emptyset$ then $l_j^{k+1} \leq l_i^k + \omega$, and $\hat{B}_j^{k+1} \subset x_i^k + B_{l_i^k + 4\omega}$. (ii) For each $j \in \mathbb{N}$ the cardinality of $\{i \in \mathbb{N} : \hat{B}_j^{k+1} \cap \hat{B}_i^k \neq \emptyset\}$ is bounded by $2L$, where $L$ is the constant in (5.5).*

PROOF. The proof of (i) follows along the lines of the proof of (iv) in Lemma 2.7. Suppose $y \in \hat{B}_j^{k+1} \cap \hat{B}_i^k = (x_j^{k+1} + B_{l_j^{k+1} + \omega}) \cap (x_i^k + B_{l_i^k + \omega})$ and $l_j^{k+1} \geq l_i^k + \omega + 1$. Then

$$x_i^k - x_j^{k+1} \in y - B_{l_i^k + \omega} - y + B_{l_j^{k+1} + \omega} \subset B_{l_j^{k+1} + 2\omega}.$$

Since $\gamma = 4\omega + 1$ and $l_i^k + \gamma \leq l_j^{k+1} + \gamma - \omega$,

$$x_i^k + B_{l_i^k + \gamma} = (x_i^k - x_j^{k+1}) + x_j^{k+1} + B_{l_i^k + \gamma} \subset x_j^{k+1} + B_{l_j^{k+1} + 2\omega} + B_{l_i^k + \gamma}$$
$$\subset x_j^{k+1} + B_{l_j + \gamma - 1},$$

by (2.11). Therefore, by (5.3) and $\Omega^{k+1} \subset \Omega^k$,

$$\emptyset \neq (x_i^k + B_{l_i^k + \gamma}) \cap (\Omega^k)^c \subset (x_j^{k+1} + B_{l_j^{k+1} + \gamma - 1}) \cap (\Omega^{k+1})^c,$$

which is a contradiction of (5.3). Hence $l_j^{k+1} \leq l_i^k + \omega$. Clearly, by (2.11)

$$\hat{B}_j^{k+1} = x_j^{k+1} + B_{l_j^{k+1}+\omega} = (x_j^{k+1} - x_i^k) + x_i^k + B_{l_j^{k+1}+\omega}$$
$$\subset x_i^k + (y - B_{l_j^{k+1}+\omega} - y + B_{l_i^k+\omega}) + B_{l_j^{k+1}+\omega}$$
$$\subset x_i^k + B_{l_j^{k+1}+2\omega} + B_{l_i^k+\omega} \subset x_i^k + B_{l_i^k+4\omega}.$$

Moreover, if $l_j^{k+1} \leq l_k^i - \omega$ then clearly $\hat{B}_j^{k+1} \subset x_i^k + B_{l_i^k+2\omega}$.

To show (ii) fix $j \in \mathbb{N}$ and consider

$$I_1 = \{i \in \mathbb{N} : \hat{B}_j^{k+1} \cap \hat{B}_i^k \neq \emptyset \quad \text{and} \quad l_i^k \geq l_j^{k+1} + \omega\}.$$

By the above $I_1 \subset \{i \in \mathbb{N} : \hat{B}_j^{k+1} \subset x_i^k + B_{l_i^k+2\omega}\} \subset \{i \in \mathbb{N} : x_j^{k+1} \in x_i^k + B_{l_i^k+2\omega}\}$ and by (5.1) and (5.5) the cardinality of $I_1$ is at most $L$. Finally, consider

$$I_2 = \{i \in \mathbb{N} : \hat{B}_j^{k+1} \cap \hat{B}_i^k \neq \emptyset \quad \text{and} \quad l_i^k \leq l_j^{k+1} + \omega\}.$$

As in the proof of (v) in Lemma 2.7, if $i \in I_2$ then

$$x_i^k + B_{l_i^k-\omega} \subset (x_i^k - x_j^{k+1}) + x_j^{k+1} + B_{l_i^k-\omega} \subset x_j^{k+1} + B_{l_i^k+\omega} + B_{l_j^{k+1}+\omega} + B_{l_i^k-\omega}$$
$$\subset x_j^{k+1} + B_{l_j^{k+1}+2\omega} + B_{l_j^{k+1}+\omega} + B_{l_j^{k+1}} \subset x_j^{k+1} + B_{l_j^{k+1}+3\omega}.$$

Since for $i \in I_2$, $x_i^k + B_{l_j^{k+1}-2\omega} \subset x_i^k + B_{l_i^k-\omega}$ we conclude by (5.2) that the cardinality of $I_2$ is less than

$$|B_{l_j^{k+1}+3\omega}|/|B_{l_j^{k+1}-2\omega}| = b^{5\omega} \leq L.$$

□

The proof of (ii) in Lemma 6.1 could be slightly shortened and simplified if we use the Whitney Lemma with $d = 6\omega + 1$ throughout Sections 5 and 6 instead of $d = 4\omega + 1$.

LEMMA 6.2. *There exists a constant $A_{10}$ independent of $i, j \in \mathbb{N}$, and $k \in \mathbb{Z}$, such that*

(6.3) $$\sup_{y \in \mathbb{R}^n} |P_{ij}^{k+1}(y)\zeta_j^{k+1}(y)| \leq A_{10} 2^{k+1}.$$

PROOF. Let $\pi_1, \ldots, \pi_m$ ($m = \dim \mathcal{P}_s$) be an orthonormal basis of $\mathcal{P}_s$ with respect to the norm

$$\|P\|^2 = \frac{1}{\int \zeta_j^{k+1}} \int |P(x)|^2 \zeta_j^{k+1}(x) dx.$$

We have

(6.4) $$P_{ij}^{k+1} = \sum_{l=1}^m \left( \frac{1}{\int \zeta_j^{k+1}} \int (f(x) - P_j^{k+1}(x))\zeta_i^k(x)\pi_l(x)\zeta_j^{k+1}(x) dx \right) \overline{\pi_l}.$$

By Lemma 5.3 applied for $\lambda = 2^{k+1}$, (5.15) and (5.16)

(6.5) $$|P_j^{k+1}(y)| \leq C 2^{k+1}, \quad |\pi_l(y)| \leq C, \quad \text{for } y \in \hat{B}_j^{k+1}.$$

Hence

(6.6) $$\left|\frac{1}{\int \zeta_j^{k+1}} \int P_j^{k+1}(y)\pi_l(y)\zeta_i^k(y)\zeta_j^{k+1}(y)dy\right| \le C2^{k+1}.$$

Therefore, by (6.4), (6.5), and (6.6) we need to show

(6.7) $$\left|\frac{1}{\int \zeta_j^{k+1}} \int f(y)\pi_l(y)\zeta_i^k(y)\zeta_j^{k+1}(y)dy\right| = |f * \Phi_{l_j^{k+1}}(w)| \le C2^{k+1},$$

for some constant $C$, where $w \in (x_j^{k+1} + B_{l_j^{k+1}+\gamma}) \cap (\Omega^{k+1})^c$, and

(6.8) $$\Phi(z) := \frac{b^{l_j^{k+1}}}{\int \zeta_j^{k+1}}(\pi_l \zeta_i^k \zeta_j^{k+1})(w - A^{l_j^{k+1}}z).$$

To see (6.7) it suffices to show that $\|\Phi\|_{\mathcal{S}_N} \le C$. However, we need only consider those values of $i$ and $j$ such that $\hat{B}_j^{k+1} \cap \hat{B}_i^k \neq \emptyset$, since otherwise $\zeta_i^k \zeta_j^{k+1}$ vanishes everywhere. Define

$$\tilde{\pi}_l(z) = \pi_l(x_j^{k+1} + A^{l_j^{k+1}}z), \quad \tilde{\zeta}_j^{k+1}(z) = \zeta_j^{k+1}(x_j^{k+1} + A^{l_j^{k+1}}z), \quad \tilde{\zeta}_i^k(z) = \zeta_i^k(w - A^{l_j^{k+1}}z).$$

Since $\operatorname{supp} \tilde{\zeta}_j^{k+1} \subset B_\omega$, by (5.15), Lemma 5.2, and the product rule, the partial derivatives of $\tilde{\pi}_l \tilde{\zeta}_j^{k+1}$ of order $\le N$ are bounded by some universal constant. Since $l_j^{k+1} \le l_i^k + \omega$ by the chain rule and Lemma 5.2, the partial derivatives of $\tilde{\zeta}_i^k$ of order $\le N$ are also bounded by some universal constant. Hence, the function $\Phi$ can be written as

$$\Phi(z) = \frac{b^{l_j^{k+1}}}{\int \zeta_j^{k+1}}(\tilde{\pi}_l \tilde{\zeta}_j^{k+1})(A^{-l_j^{k+1}}(w - x_j^{k+1}) - z)\tilde{\zeta}_i^k(z).$$

Since the above fraction is $\le b^\omega$ and $\operatorname{supp} \Phi \subset B_\gamma + B_\omega$ we can find another universal constant independent of $i, j, k, l$ so that $\|\Phi\|_{\mathcal{S}_N} \le C$. This shows (6.7). $\square$

LEMMA 6.3. *For every $k \in \mathbb{Z}$, $\sum_{i \in \mathbb{N}}(\sum_{j \in \mathbb{N}} P_{ij}^{k+1}\zeta_j^{k+1}) = 0$, where the series converges pointwise and in $\mathcal{S}'$.*

PROOF. By (5.5) the cardinality of $\{j \in \mathbb{N} : \zeta_j^{k+1}(x) \neq 0\} \le L$. Moreover, $P_{ij}^{k+1} = 0$ if $\hat{B}_j^{k+1} \cap \hat{B}_i^k = \emptyset$. By Lemma 6.1(ii) for fixed $x \in \mathbb{N}$ the series $\sum_{i \in \mathbb{N}} \sum_{j \in \mathbb{N}} P_{ij}^{k+1}(x)\zeta_j^{k+1}(x)$ contains at most $2L^2$ nonzero terms. By Lemma 6.2

(6.9) $$\sum_{i \in \mathbb{N}} \sum_{j \in \mathbb{N}} |P_{ij}^{k+1}(x)\zeta_j^{k+1}(x)| \le 2L^2 A_{10} 2^{k+1}.$$

Hence, by the Lebesgue Dominated Convergence Theorem $\sum_{i \in \mathbb{N}} \sum_{j \in \mathbb{N}} P_{ij}^{k+1}\zeta_j^{k+1}$ converges unconditionally in $\mathcal{S}'$. To conclude the proof it suffices to show that

$$\sum_{i \in \mathbb{N}} P_{ij}^{k+1} = \sum_{i \in I} P_{ij}^{k+1} = 0 \quad \text{for every } j \in \mathbb{N},$$

where $I = \{i \in \mathbb{N} : \hat{B}_j^{k+1} \cap \hat{B}_i^k \neq \emptyset\}$. Indeed, for fixed $j \in \mathbb{N}$, $\sum_{i \in \mathbb{N}} P_{ij}^{k+1}$ is an orthogonal projection of $(f - P_j^{k+1})\sum_{i \in I}\zeta_i^k$ onto $\mathcal{P}_s$ with respect to the norm (6.2). Since $\sum_{i \in I}\zeta_i^k(x) = 1$ for $x \in \hat{B}_j^{k+1}$, $\sum_{i \in \mathbb{N}} P_{ij}^{k+1}$ is an orthogonal projection of $(f - P_j^{k+1})$ with respect to the norm (6.2) which is zero by the definition of $P_j^{k+1}$. □

THEOREM 6.4 (The atomic decomposition). *If $0 < p \leq 1$, $s \geq \lfloor \ln b/(p \ln \lambda_-) \rfloor$, and $N > s$. Then $H^p \subset H^p_{\infty,s}$, where $H^p$ is given by (3.23). The inclusion map is continuous.*

PROOF. Suppose first $f \in H^p \cap L^1$. Consider the Calderón-Zygmund decomposition of $f$ of degree $s \geq \lfloor \ln b/(p \ln \lambda_-) \rfloor$ and height $2^k$ associated to $M_N f$, $f = g^k + \sum_{i \in \mathbb{N}} b_i^k$. By Lemma 5.7 (5.28), $g^k \to f$ in $H^p$ as $k \to \infty$. By Lemma 5.10(ii), $\|g^k\|_\infty \to 0$ as $k \to -\infty$. Therefore,

$$(6.10) \qquad f = \sum_{k=-\infty}^{\infty} g^{k+1} - g^k \quad \text{in } \mathcal{S}'.$$

By Lemma 6.3 and $\sum_{i \in \mathbb{N}} \zeta_i^k b_j^{k+1} = \mathbf{1}_{\Omega^k} b_j^{k+1} = b_j^{k+1}$,

$$g^{k+1} - g^k = (f - \sum_{j \in \mathbb{N}} b_j^{k+1}) - (f - \sum_{j \in \mathbb{N}} b_j^k)$$
$$= \sum_{j \in \mathbb{N}} b_j^k - \sum_{j \in \mathbb{N}} b_j^{k+1} + \sum_{i \in \mathbb{N}} (\sum_{j \in \mathbb{N}} P_{ij}^{k+1} \zeta_j^{k+1})$$
$$= \sum_{i \in \mathbb{N}} \left( b_i^k - \sum_{j \in \mathbb{N}} (\zeta_i^k b_j^{k+1} - P_{ij}^{k+1} \zeta_j^{k+1}) \right) = \sum_{i \in \mathbb{N}} h_i^k,$$

where all the series converge in $\mathcal{S}'$ and

$$h_i^k = (f - P_i^k)\zeta_i^k - \sum_{j \in \mathbb{N}} ((f - P_j^{k+1})\zeta_i^k - P_{ij}^{k+1})\zeta_j^{k+1}.$$

By the choice of $P_i^k$ and $P_{ij}^{k+1}$

$$(6.11) \qquad \int_{\mathbb{R}^n} h_i^k(x) P(x) dx = 0 \quad \text{for all } P \in \mathcal{P}_s.$$

Moreover, since $\sum_{j \in \mathbb{N}} \zeta_j^{k+1} = \mathbf{1}_{\Omega^{k+1}}$ we can write $h_i^k$ as

$$h_i^k = f \mathbf{1}_{(\Omega^{k+1})^c} \zeta_i^k - P_i^k \zeta_i^k + \sum_{j \in \mathbb{N}} P_j^{k+1} \zeta_i^k \zeta_j^{k+1} + \sum_{j \in \mathbb{N}} P_{ij}^{k+1} \zeta_j^{k+1}.$$

By Theorem 3.7, $|f(x)| \leq C_1 M_N f(x) \leq c 2^{k+1}$ for a.e. $x \in (\Omega^{k+1})^c$ and by Lemmas 5.3, 6.1, and 6.2

$$(6.12) \qquad \|h_i^k\|_\infty \leq c 2^{k+1} + A_2 2^k + 2LA_2 2^{k+1} + 2LA_{10} 2^{k+1} = C_2 2^k.$$

Recall that $P_{ij}^{k+1} \neq 0$ implies $\hat{B}_j^{k+1} \cap \hat{B}_i^k \neq 0$ and hence $\operatorname{supp} \zeta_j^{k+1} \subset \hat{B}_j^{k+1} \subset x_i^k + B_{l_i^k + 4\omega}$. Therefore

$$(6.13) \qquad \operatorname{supp} h_i^k \subset x_i^k + B_{l_i^k + 4\omega}.$$

## 6. THE ATOMIC DECOMPOSITION OF $H^p$

By (6.11), (6.12) and (6.13), $h_i^k$ is a multiple of a $(p,\infty,s)$ atom $a_i^k$, i.e., $h_i^k = \kappa_{k,i} a_i^k$, where $\kappa_{k,i} = C_2 2^k |B_{l_i^k + 4\omega}|^{1/p} = C_2 2^k b^{(l_i^k + 4\omega)/p} = C_3 2^k b^{l_i^k/p}$. By (5.2)

$$\sum_{k=-\infty}^{\infty} \sum_{i \in \mathbb{N}} (\kappa_{k,i})^p = (C_3)^p \sum_{k=-\infty}^{\infty} \sum_{i \in \mathbb{N}} 2^{kp} b^\omega |B_{i-\omega}^k| \leq (C_3)^p b^\omega \sum_{k=-\infty}^{\infty} 2^{kp} |\Omega^k|$$

(6.14)
$$\leq (C_3)^p b^\omega 2/p \sum_{k=-\infty}^{\infty} p 2^{k(p-1)} |\Omega^k| 2^{k-1}$$

$$\leq C_4 \int_0^\infty p\lambda^{p-1} |\{x : M_{N_p} f(x) > \lambda\}| d\lambda$$

$$= C_4 \|M_{N_p} f\|_p^p = C_4 \|f\|_{H^p}^p.$$

Therefore, $f = \sum_{k=-\infty}^{\infty} \sum_{i \in \mathbb{N}} h_i^k = \sum_{k=-\infty}^{\infty} \sum_{i \in \mathbb{N}} \kappa_{k,i} a_i^k$ defines an atomic decomposition of $f \in H^p \cap L^1$.

If $f$ is a general element of $H^p$ then by Corollary 5.11 we can find a sequence $\{f_m\}_{m \in \mathbb{N}} \subset H^p \cap L^1$ so that $\|f_m\|_{H^p}^p \leq 2^{2-m} \|f\|_{H^p}^p$ and $f = \sum_{m \in \mathbb{N}} f_m$. For every $m \in \mathbb{N}$ let $f_m = \sum_{i \in \mathbb{N}} \kappa_{i,m} a_m^i$ be an atomic decomposition constructed above with the summation enumerated over one index $i$. By (6.14)

$$\sum_{m \in \mathbb{N}} \sum_{i \in \mathbb{N}} (\kappa_{i,m})^p \leq C_4 \sum_{m \in \mathbb{N}} \|f_m\|_{H^p}^p \leq C_4 \|f\|_{H^p}^p \sum_{m \in \mathbb{N}} 2^{2-m} = 4C_4 \|f\|_{H^p}^p,$$

and $f = \sum_{m \in \mathbb{N}} \sum_{i \in \mathbb{N}} \kappa_{i,m} a_m^i$ defines an atomic decomposition of $f \in H^p$. □

The next theorem shows that different choices of $N \geq N_p$ in Definition 3.11 and admissible triplets $(p,q,s)$ in Definition 4.3 yield the same Hardy space $H^p$.

THEOREM 6.5 (The equivalence of norms). *Suppose $0 < p \leq 1$. If the triplet $(p,q,s)$ is admissible then $H^p = H_{q,s}^p$ and the quasi-norms $\|\cdot\|_{H^p}$ and $\|\cdot\|_{H_{q,s}^p}$ are equivalent. That is, for any $N \geq N_p$, there are constants $C_1, C_2 > 0$ so that*

(6.15) $\quad \|M_{N_p} f\|_p \leq C_1 \|f\|_{H_{q,s}^p} \leq C_1 \|f\|_{H_{\infty,s}^p} \leq C_2 \|M_N f\| \quad$ for $f \in H^p$.

PROOF. Note that if $N \geq M \geq N_p$ then $H_{(M)}^p \subset H_{(N)}^p$, where the subscript corresponds to the choice of the grand maximal function in Definition 3.11. By Definition 4.1 every $(p,\infty,s)$ atom is also a $(p,q,s)$ atom, and hence $H_{\infty,s}^p \subset H_{q,s}^p$. Therefore, if we take $N > s$ then by Theorems 6.4 and 4.5 we have

(6.16) $\quad H_{(N)}^p \subset H_{\infty,s}^p \subset H_{q,s}^p \subset H_{(N_p)}^p \subset H_{(N)}^p.$

Thus, $H_{(N)}^p = H_{(N_p)}^p = H_{q,s}^p$ for all $N \geq N_p$ and all admissible triplets $(p,q,s)$. Furthermore, since the inclusion maps in (6.16) are continuous we have (6.15). □

Theorem 6.5 encompasses fundamental properties of anisotropic Hardy spaces. As a first application of this theorem, we prove that smooth functions with certain number of vanishing moments are dense in $H^p$.

LEMMA 6.6. *Suppose $\psi \in \mathcal{S}$. For any $N \in \mathbb{N}$, there exists $C > 0$ such that for all $f \in \mathcal{S}'$ we have*

(6.17) $\quad \sup_{l \in \mathbb{Z}} M_{N+2}(f * \psi_l)(x) \leq C M_N f(x) \quad$ for all $x \in \mathbb{R}^n$.

PROOF. We claim that for any $N \in \mathbb{N}$ there exists a constant $C > 0$ so that

$$(6.18) \qquad \sup_{l \in \mathbb{Z}, l \leq 0} \|\varphi_l * \psi\|_{\mathcal{S}_N} \leq C \|\varphi\|_{\mathcal{S}_{N+2}} \|\psi\|_{\mathcal{S}_N} \qquad \text{for any } \varphi, \psi \in \mathcal{S}.$$

Indeed, take any multi-index $\alpha$, $|\alpha| \leq N$, $l \in \mathbb{Z}$, $l \leq 0$, and $x \in \mathbb{R}^n$. By (2.12)

$$\max(1, \rho(x))^N |\partial^\alpha(\varphi_l * \psi)(x)| = \max(1, \rho(x))^N |(\varphi_l * \partial^\alpha \psi)(x)|$$
$$\leq b^{\omega N} \int_{\mathbb{R}^n} \max(1, \rho(x-y))^N |\partial^\alpha \psi(x-y)| \max(1, \rho(y))^N |\varphi_l(y)| dy$$
$$\leq b^{\omega N} \|\psi\|_{\mathcal{S}_N} \int_{\mathbb{R}^n} \max(1, \rho(y))^N \max(1, \rho(A^{-l}y))^{-N-2} b^{-l} \|\varphi\|_{\mathcal{S}_{N+2}} dy$$
$$\leq b^{\omega N} \|\psi\|_{\mathcal{S}_N} \|\varphi\|_{\mathcal{S}_{N+2}} b^{-l} \int_{\mathbb{R}^n} \max(1, \rho(A^{-l}y))^{-2} dy = C \|\varphi\|_{\mathcal{S}_{N+2}} \|\psi\|_{\mathcal{S}_N}.$$

For any $f \in \mathcal{S}'$, $\psi \in \mathcal{S}$, $\varphi \in \mathcal{S}_{N+2}$, and $x \in \mathbb{R}^n$. By (6.18)

$$\sup_{k,l \in \mathbb{Z}} |(f * \psi_l) * \varphi_k(x)| \leq \sup_{\substack{k,l \in \mathbb{Z} \\ k \leq l}} |f * (\psi_l * \varphi_k)(x)| + \sup_{\substack{k,l \in \mathbb{Z} \\ k > l}} |f * (\psi_l * \varphi_k)(x)|$$
$$\leq \sup_{\substack{k,l \in \mathbb{Z} \\ k-l \leq 0}} |f * (\psi * \varphi_{k-l})_l(x)| + \sup_{\substack{k,l \in \mathbb{Z} \\ l-k < 0}} |f * (\psi_{l-k} * \varphi)_k(x)|$$
$$\leq C \|\psi\|_{\mathcal{S}_N} \|\varphi\|_{\mathcal{S}_{N+2}} M_N f(x) + C \|\psi\|_{\mathcal{S}_{N+2}} \|\varphi\|_{\mathcal{S}_N} M_N f(x)$$
$$\leq 2C \|\psi\|_{\mathcal{S}_{N+2}} M_N f(x),$$

which shows (6.17). □

As an immediate consequence of Lemma 6.6 we obtain

COROLLARY 6.7. *If $f \in H^p$ and $\varphi \in \mathcal{S}$ then $f * \varphi \in H^p$.*

A less immediate consequence of Lemma 6.6 is

THEOREM 6.8. *Suppose $\psi \in \mathcal{S}$ and $\int \psi = 1$. For any $0 < p < \infty$ and $f \in H^p$ we have*

$$(6.19) \qquad f * \psi_l \to f \quad \text{in } H^p \text{ as } l \to -\infty.$$

PROOF. Suppose first that $f \in L^1 \cap H^p$. By Lemma 6.6 and the Lebesgue Dominated Convergence Theorem it suffices to show

$$(6.20) \qquad M_N(f * \psi_l - f)(x) \to 0 \qquad \text{for a.e. } x \in \mathbb{R}^n \text{ as } l \to -\infty,$$

where $N = N_p + 2$. Note that if $g$ is continuous with compact support then

$$\|M_N(g * \psi_l - g)\|_\infty \leq C \|g * \psi_l - g\|_\infty \to 0 \qquad \text{as } l \to -\infty.$$

For any $\varepsilon > 0$ we can find a continuous function $g$ with compact support such that $\|f - g\|_1 < \varepsilon$. By Lemma 6.6

$$\limsup_{l \to -\infty} M_N(f * \psi_l - f)(x)$$
$$\leq \sup_{l \in \mathbb{Z}} M_N((f-g) * \psi_l)(x) + \limsup_{l \to -\infty} M_N(g * \psi_l - g)(x) + M_N(f-g)(x)$$
$$\leq C M_{N_p}(f-g)(x)$$

By Theorem 3.6 for any $\lambda > 0$

$$|\{x : \limsup_{l \to -\infty} M_N(f * \psi_l - f)(x) > \lambda\}|$$
$$\leq |\{x : M_{N_p}(f-g)(x) > \lambda/C\}| \leq C'\|f-g\|_1/\lambda < C'\varepsilon/\lambda.$$

Since $\varepsilon > 0$ is arbitrary we have (6.20) for every $f \in L^1 \cap H^p$.

Recall that $L^1 \cap H^p$ is dense in $H^p$. This follows from Corollary 5.11 ($0 < p \leq 1$) and the fact that $L^1 \cap L^p$ is dense in $L^p$ ($1 < p < \infty$). Alternatively, we could use the fact that finite linear combinations of atoms are dense in $H^p$ by Theorem 6.4. Moreover, by Lemma 6.6 the convolution operators $R_l(f) = f * \psi_l$ are uniformly bounded on $H^p$ for all $l \in \mathbb{Z}$.

Hence, if $f$ is an arbitrary element of $H^p$ then for every $\varepsilon > 0$ we can find $g \in L^1 \cap H^p$ such that $\|f-g\|_{H^p}^p < \varepsilon$. Therefore,

$$\limsup_{l \to -\infty} \|f * \psi_l - f\|_{H^p}^p$$
$$\leq \sup_{l \in \mathbb{Z}} \|(f-g) * \psi_l\|_{H^p}^p + \|f-g\|_{H^p}^p + \limsup_{l \to -\infty} \|g * \psi_l - g\|_{H^p}^p$$
$$\leq C\|f-g\|_{H^p}^p \leq C\varepsilon.$$

Since $\varepsilon > 0$ was arbitrary we obtain (6.19). □

Finally, we prove that smooth functions with compact support are dense in $H^p$.

THEOREM 6.9. $H^p \cap C_0^\infty$ is dense in $H^p$, where $C_0^\infty$ denotes $C_0^\infty$ functions with compact support.

PROOF. Suppose $f$ is a finite linear combination of atoms. Take $\psi \in C_0^\infty$. Clearly, $f * \psi_l \in C_0^\infty$ for every $l \in \mathbb{Z}$ and $f * \psi_l \to f$ in $H^p$ as $l \to -\infty$ by Theorem 6.8. This finishes the proof since finite linear combinations of atoms are dense in $H^p$. □

REMARK. $C_0^\infty$ is **not** a subset of $H^p$ for $p \leq 1$. It is not hard to see that $\psi \in C_0^\infty$ with $\int \psi \neq 0$ does not belong to any $H^p$ for $p \leq 1$. However, if $\psi \in C_0^\infty$ has vanishing moments of order $s$, i.e., $\int \psi P = 0$ for all $P \in \mathcal{P}_s$, where $s \geq \lfloor (1/p - 1) \ln b / \ln \lambda_- \rfloor$, then $\psi$ is a constant multiple of a $(p, \infty, s)$ atom and hence $\psi \in H^p$.

## 7. Other maximal definitions

The goal of this section is to prove the characterization of Hardy spaces using the radial and nontangential maximal functions of a single test function $\varphi \in \mathcal{S}$, $\int \varphi \neq 0$. This result was first shown for the classical isotropic $H^p$ spaces by C. Fefferman and Stein in their fundamental work [FS2, St2, Chapter III]. Analogous results were shown by Calderón and Torchinsky [CT1, CT2] for parabolic $H^p$ spaces, Folland and Stein [FoS] for $H^p$ spaces on homogeneous groups, and Uchyiama [Uc1] for $H^p$ on the space of homogeneous type.

Theorem 7.1 extends this result of C. Fefferman and Stein to the setting of anisotropic $H^p$ spaces introduced in Chapter 3.

THEOREM 7.1. *Suppose $\varphi \in \mathcal{S}$, $\int \varphi \neq 0$, and $0 < p < \infty$. Then for any $f \in \mathcal{S}'$ the following are equivalent:*

(7.1) $$f \in H^p,$$

(7.2) $$M_\varphi f \in L^p,$$

(7.3) $$M_\varphi^0 f \in L^p.$$

*In this case*
$$||f||_{H^p} = ||M_N f||_p \leq C_1 ||M_\varphi f||_p \leq C_2 ||M_\varphi^0 f||_p,$$
*for sufficiently large $N$, $C_1$, $C_2$ independent of $f \in H^p$.*

We start with a very useful result about maximal functions with different apertures. Suppose $F: \mathbb{R}^n \times \mathbb{Z} \to [0, \infty)$ is an arbitrary (possibly nonmeasurable) function. In our cases $F$ is going to be (at least) Borel measurable. For fixed $l \in \mathbb{Z}$ and $K \in \mathbb{Z} \cup \{\infty\}$ define the maximal function of $F$ with *aperture $l$* as

(7.4) $$F_l^*(x) = F_l^{*K}(x) := \sup_{\substack{k \in \mathbb{Z} \\ k \leq K}} \sup_{y \in x + B_{k+l}} F(y, k).$$

Note that $F_l^*: \mathbb{R}^n \to [0, \infty]$ is lower semicontinuous, i.e., $\{x : F_l^*(x) > \lambda\}$ is open for all $\lambda > 0$. Indeed, if $F_l^*(x) > \lambda$ for some $x \in \mathbb{R}^n$, i.e., there is $k \leq K$ and $y \in x + B_{k+l}$ so that $F(y, k) > \lambda$, then $y \in x' + B_{k+l}$ for $x'$ in a sufficiently small neighborhood of $x$. Thus $F_l^*(x') > \lambda$. Here we need that the balls $B_j$ are open for all $j \in \mathbb{Z}$.

LEMMA 7.2. *There exists a constant $C > 0$ so that for all functions $F: \mathbb{R}^n \times \mathbb{Z} \to [0, \infty)$, $\lambda > 0$, $l \geq l' \in \mathbb{Z}$, and $K \in \mathbb{Z} \cup \{\infty\}$ we have*

(7.5) $$|\{x : F_l^{*K}(x) > \lambda\}| \leq C b^{l-l'} |\{x : F_{l'}^{*K}(x) > \lambda\}|.$$

*In particular,*

(7.6) $$\int_{\mathbb{R}^n} F_l^{*K}(x) dx \leq C b^{l-l'} \int_{\mathbb{R}^n} F_{l'}^{*K}(x) dx.$$

PROOF. Let $\Omega = \{x : F_{l'}^*(x) > \lambda\}$. Suppose $F_l^*(x) > \lambda$ for some $x \in \mathbb{R}^n$. Then there exist $k \leq K$, $y \in x + B_{k+l}$ such that $F(y, k) > \lambda$. Clearly, $y + B_{k+l'} \subset \Omega$. Also $y + B_{k+l'} \subset x + B_{k+l} + B_{k+l'} \subset x + B_{k+l+\omega}$. Hence $y + B_{k+l'} \subset \Omega \cap (x + B_{k+l+\omega})$ and
$$|\Omega \cap (x + B_{k+l+\omega})| \geq b^{k+l'} = b^{k+l+\omega} b^{l'-l-\omega}.$$

Therefore, $M_{HL}(\mathbf{1}_\Omega)(x) \geq b^{l'-l-\omega}$, where $M_{HL}$ is the centered Hardy-Littlewood maximal function $M_{HL}f(y) := \sup_{m\in\mathbb{Z}} b^{-m}\int_{y+B_m}|f(z)|dz$. By Theorem 3.6

$$|\{x : F_l^*(x) > \lambda\}| \leq |\{x : M_{HL}(\mathbf{1}_\Omega)(x) \geq b^{l'-l-\omega}\}| \leq Cb^{l-l'+\omega}\|\mathbf{1}_\Omega\|_1 = C'b^{l-l'}|\Omega|,$$

hence (7.5) holds. Integrating (7.5) on $(0,\infty)$ with respect to $\lambda$ yields (7.6). □

The following result enables us to pass from one function in $\mathcal{S}$ to the sum of negative dilates of another function in $\mathcal{S}$ with nonzero mean.

LEMMA 7.3. *Suppose $\varphi \in \mathcal{S}$ with $\int \varphi \neq 0$. For every $\psi \in \mathcal{S}$ there is a sequence of test functions $(\eta^j)_{j=0}^\infty$, $\eta^j \in \mathcal{S}$ such that*

$$(7.7) \qquad \psi = \sum_{j=0}^\infty \eta^j * \varphi_{-j},$$

*where the convergence is in $\mathcal{S}$.*

*Furthermore, for any integers $L, N > 0$ there exist a constant $C$ and an integer $M > 0$ (depending on $L$ and $N$, but independent of the choice of $\psi$) such that*

$$(7.8) \qquad \|\eta^j\|_{\mathcal{S}_N} \leq Cb^{-jL}\|\psi\|_{\mathcal{S}_M} \qquad \text{for all } j \geq 0.$$

PROOF. For the purpose of the proof we will work with the standard definition of $\mathcal{S}_N$, that is

$$\|\eta\|_{\mathcal{S}_N} = \sup_{x\in\mathbb{R}^n}\sup_{|\alpha|\leq N}(1+|x|)^N|\partial^\alpha\eta(x)|.$$

By Lemma 3.2 this implies the corresponding result for the usual homogeneous $\mathcal{S}_N$. We claim that for every integer $N > 0$ there is a constant $C$ so that

$$(7.9) \qquad \|\hat{\eta}\|_{\mathcal{S}_N} \leq C\|\eta\|_{\mathcal{S}_{N+n+1}}.$$

Indeed, for any multi-indices $|\alpha|, |\beta| \leq N$ we have

$$(2\pi i)^{|\beta|}\xi^\beta \partial^\alpha \hat{\eta}(\xi) = (-2\pi i)^{|\alpha|}\int_{\mathbb{R}^n} e^{-2\pi i\langle x,\xi\rangle}\partial^\beta(x^\alpha\eta(x))dx.$$

Hence, by multiplying and dividing the right hand side by $(1+|x|)^{n+1}$ we have

$$|\xi^\beta \partial^\alpha \hat{\eta}(\xi)| \leq (2\pi)^{|\alpha|-|\beta|}\sup_{x\in\mathbb{R}^n}(1+|x|)^{n+1}|\partial^\beta(x^\alpha\eta(x))|\int_{\mathbb{R}^n}(1+|x|)^{-n-1}dx,$$

which implies (7.9).

Let $\hat{\Delta}$ be an expansive ellipsoid for the dilation $B = A^T$, i.e., $\hat{\Delta} \subset r\hat{\Delta} \subset B\hat{\Delta}$, guaranteed by Lemma 2.2. By scaling of $\varphi$ and $\hat{\Delta}$ we can assume that $\int \varphi = 1$ and $|\hat{\varphi}(\xi)| \geq 1/2$ for $\xi \in B\hat{\Delta}$. Consider a $C^\infty$ function $\zeta$ such that $\zeta \equiv 1$ on $\hat{\Delta}$ and $\operatorname{supp}\zeta \subset B\hat{\Delta}$. Define a sequence of functions $(\zeta_j)_{j=0}^\infty$ by $\zeta_0 = \zeta$,

$$\zeta_j(\xi) = \zeta(B^{-j}\xi) - \zeta(B^{-j+1}\xi) \qquad \text{for } j \geq 1.$$

Clearly,

$$\sum_{j=0}^\infty \zeta_j(\xi) = 1 \qquad \text{for all } \xi \in \mathbb{R}^n.$$

Thus
$$\hat{\psi}(\xi) = \sum_{j=0}^{\infty} \zeta_j(\xi)\hat{\psi}(\xi) = \sum_{j=0}^{\infty} \frac{\zeta_j(\xi)}{\hat{\varphi}(B^{-j}\xi)}\hat{\psi}(\xi)\hat{\varphi}(B^{-j}\xi).$$

For $j \geq 0$ define $\eta^j$ by

(7.10) $$\hat{\eta}^j(\xi) = \frac{\zeta_j(\xi)}{\hat{\varphi}(B^{-j}\xi)}\hat{\psi}(\xi).$$

Then, by the choice of $\eta^j$'s, (7.7) holds. Moreover, the convergence in $\mathcal{S}$ of the series (7.7) is a direct consequence of (6.18) and (7.8). Therefore, it remains to show (7.8).

To show (7.8), take $M = N + 2(n+1) + L\lceil \ln b / \ln \lambda_- \rceil$, where $\lambda_-$ is the same as in (2.1). Our goal is to show that there exists a constant $C$ (independent of $\psi \in \mathcal{S}$) such that
$$\|\hat{\eta}^j\|_{\mathcal{S}_{N+n+1}} \leq Cb^{-jL}\|\psi\|_{\mathcal{S}_M} \quad \text{for } j \geq 0.$$

Let $\tilde{\zeta}(\xi) = (\zeta(\xi) - \zeta(B\xi))/\hat{\varphi}(\xi)$. Since $\operatorname{supp}\tilde{\zeta} \subset B\hat{\Delta}$ and $|\hat{\varphi}(\xi)| \geq 1/2$ for $\xi \in B\hat{\Delta}$ we have $\sup_\xi \sup_{|\alpha| \leq N+n+1} |\partial^\alpha \tilde{\zeta}(\xi)| \leq C$ for some constant $C$. Since $\tilde{\zeta}(B^{-j}\xi) = \zeta_j(\xi)/\hat{\varphi}(B^{-j}\xi)$ by the chain rule we have
$$\sup_\xi \sup_{|\alpha| \leq N+n+1} |\partial^\alpha(\zeta_j(\cdot)/\hat{\varphi}(B^{-j}\cdot))(\xi)| \leq C.$$

By (7.10), the product rule and since $\zeta_j(\xi) = 0$ for $\xi \in B^{j-1}\hat{\Delta}$

$$\|\hat{\eta}^j\|_{\mathcal{S}_{N+n+1}} \leq C \sup_{\xi \notin B^{j-1}\hat{\Delta}} \sup_{|\alpha| \leq N+n+1} (1+|\xi|)^{N+n+1} |\partial^\alpha \hat{\psi}(\xi)|$$

$$\leq C \sup_{\xi \notin B^{j-1}\hat{\Delta}} (1+|\xi|)^{-L\lceil \ln b/\ln \lambda_-\rceil} \sup_{|\alpha| \leq N+n+1} (1+|\xi|)^{N+n+1+L\lceil \ln b/\ln \lambda_-\rceil} |\partial^\alpha \hat{\psi}(\xi)|$$

$$\leq Cb^{-jL}\|\hat{\psi}\|_{\mathcal{S}_{M-n-1}} \leq Cb^{-jL}\|\psi\|_{\mathcal{S}_M},$$

because for $\xi \notin \hat{\Delta}$, $(1+|B^{j-1}\xi|) \geq 1/c\lambda_-^{j-1} \inf_{\xi \notin \hat{\Delta}} |\xi|$. This finishes the proof of (7.8). Since our choice of $N$ is arbitrary in (7.8) the series in (7.7) converges in $\mathcal{S}$. □

We now introduce maximal functions obtained from truncation with an additional extra decay term. Namely, for an integer $K$ representing the truncation level and real number $L \geq 0$ representing the decay level, we define radial, nontangential, tangential, radial grand, and nontangential grand maximal functions, respectively as

$$M_\varphi^{0(K,L)}f(x) = \sup_{\substack{k \in \mathbb{Z} \\ k \leq K}} |(f * \varphi_k)(x)| \max(1, \rho(A^{-K}x))^{-L}(1+b^{-k-K})^{-L},$$

$$M_\varphi^{(K,L)}f(x) = \sup_{\substack{k \in \mathbb{Z} \\ k \leq K}} \sup_{y \in x+B_k} |(f * \varphi_k)(y)| \max(1, \rho(A^{-K}y))^{-L}(1+b^{-k-K})^{-L},$$

$$T_\varphi^{N(K,L)}f(x) = \sup_{\substack{k \in \mathbb{Z} \\ k \leq K}} \sup_{y \in \mathbb{R}^n} \frac{|(f*\varphi_k)(y)|}{\max(1,\rho(A^{-k}(x-y)))^N} \frac{(1+b^{-k-K})^{-L}}{\max(1,\rho(A^{-K}y))^L},$$

$$M_N^{0(K,L)}f(x) = \sup_{\varphi \in \mathcal{S}_N} M_\varphi^{0(K,L)}f(x).$$

$$M_N^{(K,L)}f(x) = \sup_{\varphi \in \mathcal{S}_N} M_\varphi^{(K,L)}f(x).$$

The next lemma guarantees the control of the tangential by the nontangential maximal function independent of $K$ and $L$.

LEMMA 7.4. *Suppose $p > 0$, $N > 1/p$, and $\varphi \in \mathcal{S}$. Then there exists a constant $C$ so that for all $K \in \mathbb{Z}$, $L \geq 0$, and $f \in \mathcal{S}'$,*

$$\|T_\varphi^{N(K,L)} f\|_p \leq C \|M_\varphi^{(K,L)} f\|_p. \tag{7.11}$$

PROOF. Consider function $F : \mathbb{R}^n \times \mathbb{Z} \to [0, \infty)$ given by

$$F(y, k) = |(f * \varphi_k)(y)|^p \max(1, \rho(A^{-K} y))^{-pL} (1 + b^{-k-K})^{-pL}.$$

Fix $x \in \mathbb{R}^n$. If $k \leq K$ and $x - y \in B_{k+1}$ then

$$F(y, k) \max(1, \rho(A^{-k}(x - y)))^{-pN} \leq F_0^{*K}(x).$$

If $x - y \in B_{k+j+1} \setminus B_{k+j}$ for some $j \geq 1$ then

$$F(y, k) \max(1, \rho(A^{-k}(x - y)))^{-pN} \leq F_j^{*K}(x) b^{-jNp}.$$

By taking supremum over all $y \in \mathbb{R}^n$, $k \leq K$ we have

$$(T_\varphi^{N(K,L)} f(x))^p \leq \sum_{j=0}^\infty F_j^{*K}(x) b^{-jNp}.$$

Therefore by Lemma 7.2,

$$\|T_\varphi^{N(K,L)} f\|_p^p \leq \sum_{j=0}^\infty b^{-jNp} \int_{\mathbb{R}^n} F_j^{*K}(x) dx$$

$$\leq C \sum_{j=0}^\infty b^{-jNp} b^j \int_{\mathbb{R}^n} F_0^{*K}(x) dx \leq C' \|M_\varphi^{(K,L)} f\|_p^p,$$

where $C' = C \sum_{j=0}^\infty b^{j(1-Np)} < \infty$. □

Lemma 7.5 gives the pointwise majorization of the grand maximal function by the tangential one.

LEMMA 7.5. *Suppose $\varphi \in \mathcal{S}$ and $\int \varphi \neq 0$. For given $N, L \geq 0$ there exists an integer $M > 0$ and a constant $C > 0$ so that for all $f \in \mathcal{S}'$ and integers $K \geq 0$ we have*

$$M_M^{0(K,L)} f(x) \leq C T_\varphi^{N(K,L)} f(x) \quad \text{for all } x \in \mathbb{R}^n. \tag{7.12}$$

PROOF. Take any $\psi \in \mathcal{S}$. By Lemma 7.3 there exists a sequence of test functions $(\eta^j)_{j=0}^\infty$ so that (7.7) holds. For a fixed integer $k \leq K$ and $x \in \mathbb{R}^n$,

$$|(f * \psi_k)(x)| = \left|\left(f * \sum_{j=0}^\infty (\eta^j * \varphi_{-j})_k\right)(x)\right| = \left|\left(f * \sum_{j=0}^\infty (\eta^j)_k * \varphi_{-j+k}\right)(x)\right|$$

$$\leq \sum_{j=0}^\infty |(f * (\eta^j)_k * \varphi_{-j+k})(x)| \leq \sum_{j=0}^\infty \int_{\mathbb{R}^n} |(f * \varphi_{-j+k})(x-y)||(\eta^j)_k(y)| dy$$

$$\leq T_\varphi^{N(K,L)} f(x)$$

$$\cdot \sum_{j=0}^\infty \int_{\mathbb{R}^n} \max(1, \rho(A^{j-k} y))^N \max(1, \rho(A^{-K}(x-y)))^L (1 + b^{j-k-K})^L |(\eta^j)_k(y)| dy.$$

Therefore

$$M_\psi^{0(K,L)} f(x) \leq T_\varphi^{N(K,L)} f(x) \sup_{\substack{k \in \mathbb{Z} \\ k \leq K}} \sum_{j=0}^\infty \int_{\mathbb{R}^n} \max(1, \rho(A^{j-k} y))^N$$

$$\frac{\max(1, \rho(A^{-K}(x-y)))^L (1 + b^{j-k-K})^L}{\max(1, \rho(A^{-K} x))^L (1 + b^{-k-K})^L} |(\eta^j)_k(y)| dy.$$

Using the inequalities (2.12), (2.13) and the change of variables we can estimate the last sum by

$$\sum_{j=0}^\infty \int_{\mathbb{R}^n} \max(1, \rho(A^{j-k} y))^N b^{\omega L} \max(1, \rho(A^{-K} y))^L 2^L b^{jL} b^{-k} |\eta^j(A^{-k} y)| dy$$

$$\leq 2^L b^{\omega L} \sum_{j=0}^\infty b^{j(N+L)} \int_{\mathbb{R}^n} \max(1, \rho(y))^{N+L} |\eta^j(y)| dy \leq 2^L b^{\omega L} \sum_{j=0}^\infty b^{j(N+L)} \|\eta^j\|_{\mathcal{S}_{N+L}}.$$

By Lemma 7.3 there exists an integer $M \geq 0$ so that

$$\|\eta^j\|_{\mathcal{S}_{N+L}} \leq C b^{-j(N+L+1)} \|\psi\|_{\mathcal{S}_M},$$

and hence

$$M_M^{0(K,L)} f(x) = \sup_{\psi \in \mathcal{S}_M} M_\psi^{0(K,L)} f(x) \leq 2^L b^{\omega L} \sum_{j=0}^\infty C b^{-j} T_\varphi^{N(K,L)} f(x)$$

$$= C' T_\varphi^{N(K,L)} f(x).$$

This shows (7.12). □

LEMMA 7.6. *Suppose $p > 0$, $\varphi \in \mathcal{S}$, and $f \in \mathcal{S}'$. Then for every $M > 0$ there exists $L > 0$ so that*

(7.13) $\qquad M_\varphi^{(K,L)} f(x) \leq C \max(1, \rho(x))^{-M} \qquad$ *for all $x \in \mathbb{R}^n$,*

*for some constant $C = C(K)$ dependent on $K \geq 0$.*

PROOF. There exists an integer $N > 0$ and a constant $C > 0$ so that

$$|f * \varphi(x)| \leq C\|\varphi\|_{\mathcal{S}_N} \max(1, \rho(x))^N \qquad \text{for all } x \in \mathbb{R}^n, \varphi \in \mathcal{S}.$$

If $k \leq 0$ then by the chain rule

$$\|\varphi_k\|_{\mathcal{S}_N} = \sup_{z \in \mathbb{R}^n} \sup_{|\alpha| \leq N} \max(1, \rho(z))^N b^{-k} |\partial^\alpha \varphi(A^{-k} \cdot)(z)|$$

$$\leq b^{-k} \sup_{z \in \mathbb{R}^n} \sup_{|\alpha| \leq N} \max(1, \rho(z))^N |\partial^\alpha \varphi(A^{-k} z)| C \|A^{-k}\|^N \leq C b^{-k} \lambda_+^{-kN} \|\varphi\|_{\mathcal{S}_N}.$$

If $0 < k \leq K$ then by the chain rule

$$\|\varphi_k\|_{\mathcal{S}_N} \leq C \sup_{z \in \mathbb{R}^n} \sup_{|\alpha| \leq N} \max(1, \rho(z))^N |\partial^\alpha \varphi(A^{-k} z)| \leq C b^{KN} \|\varphi\|_{\mathcal{S}_N}.$$

Therefore, if we take $L = N + M$, then for any $k \leq K$ and $y \in x + B_k$,

$$|f * \varphi_k(y)| \max(1, \rho(A^{-K} y))^{-L} (1 + b^{-k-K})^{-L}$$
$$\leq C \max(1, \rho(y))^N \|\varphi_k\|_{\mathcal{S}_N} b^{KL} \max(1, \rho(y))^{-L} b^{(k+K)L}$$
$$\leq C \max(1, \rho(y))^{N-L} b^{3KL+KN} \|\varphi\|_{\mathcal{S}_N}$$
$$\leq C b^{(K+\omega)M} b^{3KL+KN} \max(1, \rho(x))^{-M} \|\varphi\|_{\mathcal{S}_N}.$$

This shows (7.13) and finishes the proof of Lemma 7.6. $\square$

Note that the above argument gives the same estimate for the truncated grand maximal function $M_N^{0(K,L)} f$. As a consequence of Lemma 7.6 we have that for any choice of $K \geq 0$ and any $f \in \mathcal{S}'$ we can find an appropriate $L > 0$ so that the maximal function, say $M_\varphi^{(K,L)} f$, is bounded and belongs to $L^p(\mathbb{R}^n)$. This becomes crucial in the proof of Theorem 7.1, where we work with truncated maximal functions. The complexity of the preceding argument stems from the fact that a priori we do not know whether $M_\varphi^0 f \in L^p$ implies $M_\varphi f \in L^p$. Instead we must work with variants of maximal functions for which this is satisfied.

PROOF OF THEOREM 7.1. Clearly, (7.1) $\implies$ (7.2) $\implies$ (7.3). By Lemma 7.5 applied for $N > 1/p$ and $L = 0$ we have pointwise estimate $M_M^{0(K,0)} f(x) \leq C T^{N(K,0)} f(x)$ for all $f \in \mathcal{S}'$ and integers $K \geq 0$. By Lemma 7.4 we have another constant $C$ so that

$$\|M_M^{0(K,0)} f\|_p \leq C \|M_\varphi^{(K,0)} f\|_p \qquad \text{for } f \in \mathcal{S}', K \geq 0.$$

As $K \to \infty$ we obtain $\|M_M^0 f\|_p \leq C \|M_\varphi f\|_p$ by the Monotone Convergence Theorem. It remains to show (7.3) $\implies$ (7.2).

Suppose now $M_\varphi^0 f \in L^p$. By Lemma 7.6 we can find $L > 0$ so that (7.13) holds, i.e., $M_\varphi^{(K,L)} f \in L^p$ for all $K \geq 0$. By Lemmas 7.4 and 7.5 we can find $M > 0$ so that $\|M_M^{0(K,L)} f\|_p \leq C_1 \|M_\varphi^{(K,L)} f\|_p$ with a constant $C_1$ independent of $K \geq 0$. For given $K \geq 0$ let

(7.14) $\qquad \Omega_K = \{x \in \mathbb{R}^n : M_M^{0(K,L)} f(x) \leq C_2 M_\varphi^{(K,L)} f(x)\},$

where $C_2 = 2^{1/p} C_1$. We claim that

(7.15) $\qquad \int_{\mathbb{R}^n} M_\varphi^{(K,L)} f(x)^p dx \leq 2 \int_{\Omega_K} M_\varphi^{(K,L)} f(x)^p dx.$

Indeed, this follows from

$$\int_{\Omega_K^c} M_\varphi^{(K,L)} f(x)^p dx \le C_2^{-p} \int_{\Omega_K^c} M_M^{(K,L)} f(x)^p dx \le (C_1/C_2)^p \int_{\mathbb{R}^n} M_\varphi^{(K,L)} f(x)^p dx,$$

and $(C_1/C_2)^p = 1/2$.

We also claim that for $0 < q < p$ there exists $C_3 > 0$ so that for all integers $K \ge 0$

(7.16) $\qquad M_\varphi^{(K,L)} f(x) \le C_3 (M_{HL}(M_\varphi^{0(K,L)} f(\cdot)^q)(x))^{1/q} \qquad$ for $x \in \Omega_K$.

Indeed, let

$$F(y,k) = |(f * \varphi_k)(y)| \max(1, \rho(A^{-K}y))^{-L} (1 + b^{-k-K})^{-L}.$$

Suppose $x \in \Omega_K$. There is $k \le K$ and $y \in x + B_k$ such that

$$F(y,k) \ge F_0^*(x)/2 = M_\varphi^{(K,L)} f(x)/2.$$

Consider $x' \in y + B_{k-l}$ for some integer $l \ge 0$ to be specified later. We have

$$f * \varphi_k(x') - f * \varphi_k(y) = f * \Phi_k(y), \qquad \text{where} \quad \Phi(z) = \varphi(z + A^{-k}(x'-y)) - \varphi(z).$$

By the Mean Value Theorem

$$\|\Phi\|_{S_M} \le \sup_{h \in B_{-l}} \|\varphi(\cdot + h) - \varphi(\cdot)\|_{S_M}$$

$$= \sup_{h \in B_{-l}} \sup_{z \in \mathbb{R}^n} \sup_{|\alpha| \le M} \max(1, \rho(z)^M) |\partial^\alpha \varphi(z+h) - \partial^\alpha \varphi(z)|$$

$$\le C \sup_{h \in B_{-l}} \sup_{z \in \mathbb{R}^n} \sup_{|\alpha| \le M+1} \max(1, \rho(z+h)^M) |\partial^\alpha \varphi(z+h)| \cdot \sup_{h \in B_{-l}} |h| \le C_4 \lambda_-^{-l},$$

where $C_4$ does not depend on $L$. Since $\max(1, \rho(A^{-K}x')) \le b^\omega \max(1, \rho(A^{-K}y))$ we have

$$b^{\omega L} F(x', k) \ge (|f * \varphi_k(y)| - |f * \Phi_k(y)|) \max(1, \rho(A^{-K}y))^{-L} (1 + b^{-k-K})^{-L}$$
$$\ge F(y,k) - M_M^{(K,L)} f(x) \|\Phi\|_{S_M} \ge M_\varphi^{(K,L)} f(x)/2 - b^{\omega M} M_M^{0(K,L)} f(x) C_4 \lambda_-^{-l}$$
$$\ge M_\varphi^{(K,L)} f(x)/2 - b^{\omega M} C_4 \lambda_-^{-l} C_2 M_\varphi^{(K,L)} f(x) \ge M_\varphi^{(K,L)} f(x)/4,$$

if we choose for $l$ the smallest integer $l \ge 0$ so that $b^{\omega M} C_4 \lambda_-^{-l} C_2 \le 1/4$. Here we used the fact that $x \in \Omega_K$ and the pointwise majorization of nontangential by radial grand maximal function,

$$M_M^{(K,L)} f(x) \le b^{\omega M} M_M^{0(K,L)} f(x),$$

see Proposition 3.10. Therefore for $x \in \Omega_K$,

$$M_\varphi^{(K,L)} f(x)^q \le \frac{4^q b^{\omega Lq}}{|B_{k-l}|} \int_{y+B_{k-l}} F(z,k)^q dz$$

$$\le 4^q b^{\omega Lq} \frac{b^{\omega+l}}{|B_{k+\omega}|} \int_{x+B_{k+\omega}} M_\varphi^{0(K,L)} f(z)^q dz \le C_3 M_{HL}(M_\varphi^{0(K,L)} f(\cdot)^q)(x),$$

which shows (7.16). Finally, by (7.15), (7.16), and the Maximal Theorem 3.6

$$
\begin{aligned}
(7.17) \quad \int_{\mathbb{R}^n} M_\varphi^{(K,L)} f(x)^p dx &\leq 2 \int_{\Omega_K} M_\varphi^{(K,L)} f(x)^p dx \\
&\leq 2C_3^{p/q} \int_{\Omega_K} M_{HL}(M_\varphi^{0(K,L)} f(\cdot)^q)(x)^{p/q} dx \leq C_5 \int_{\mathbb{R}^n} M_\varphi^{0(K,L)} f(x)^p dx,
\end{aligned}
$$

where the constant $C_5$ depends on $p/q > 1$ and $L \geq 0$, but is independent of $K \geq 0$. This inequality is crucial as it gives a bound of nontangential by radial maximal function in $L^p$. The rest of the proof is immediate.

Since $M_\varphi^{(K,L)} f(x)$ converges pointwise and monotonically to $M_\varphi f(x)$ for all $x \in \mathbb{R}^n$ as $K \to \infty$, $M_\varphi f \in L^p(\mathbb{R}^n)$ by (7.17) and the Monotone Convergence Theorem. Therefore, we can now choose $L = 0$ and again by (7.17) and the Monotone Convergence Theorem we have $\|M_\varphi f\|_p^p \leq C_5 \|M_\varphi^0 f\|_p^p$, where now $C_5$ corresponds to $L = 0$ and is independent of $f \in \mathcal{S}'$. This concludes the proof of Theorem 7.1. □

REMARK. Theorem 7.1 shows the equivalence of $L^p$ norms of radial, nontangental, and grand maximal functions of a tempered distribution $f \in \mathcal{S}'$. One could ask if a similar result holds for general (not necessarily tempered) distributions $f \in \mathcal{D}'(\mathbb{R}^n)$. Naturally, in this case it is necessary to assume that $\varphi$ is an element of the space $\mathcal{D}(\mathbb{R}^n)$ of compactly supported and $C^\infty$ test functions, and $\int \varphi \neq 0$. Uchiyama showed that this is indeed true for isotropic $H^p$ spaces and the proof of this very non-trivial result can be found in [Uc2, Uc3, Chapter IX].

## 8. Duals of $H^p$

In this section we provide the description of the duals of anisotropic $H^p$ spaces, $0 < p \leq 1$, in terms of Campanato spaces. This description is a consequence of the atomic decomposition of $H^p$ and some functional analysis and approximation arguments.

The dual of $H^p$ space of holomorphic functions on the unit disc for $0 < p < 1$ was first identified with a Lipshitz space by Duren, Romberg, and Shields [DRS]. For the classical Stein-Weiss $H^p$ space on $\mathbb{R}^n$ this result is due to Walsh [Wa]. The identification of the dual of $H^1$ as the space of $BMO$ is a famous result of C. Fefferman [Fe, FS2]. Characterizations of $(H^p)^*$ in terms of BMO, Campanato, and Lipschitz spaces in various setting, other than isotropic $H^p(\mathbb{R}^n)$, were obtained by many other authors; for example, the dual of parabolic $H^p$ spaces was determined by Calderón and Torchinsky [CT2].

We start with the definition of Campanato spaces [Cm] which are defined in terms of approximations by polynomials on dilated balls, generalizing the BMO (bounded mean oscillation) space introduced by John and Nirenberg [JN].

DEFINITION 8.1. Let $\mathcal{B}$ denote the collection of dilated balls associated to the dilation $A$, i.e., $\mathcal{B} = \{x + B_k : x \in \mathbb{R}^n, k \in \mathbb{Z}\}$. If $l \geq 0$, $1 \leq q \leq \infty$, and $s = 0, 1, \ldots$, we define the *Campanato space* $C_{q,s}^l$ to be the space of all locally $L^q$ functions $g$ on $\mathbb{R}^n$ so that

$$(8.1) \quad \|g\|_{C_{q,s}^l} := \sup_{B \in \mathcal{B}} \inf_{P \in \mathcal{P}_s} |B|^{-l} \left( \frac{1}{|B|} \int_B |g(x) - P(x)|^q dx \right)^{1/q} < \infty, \qquad (q < \infty)$$

$$(8.2) \quad \|g\|_{C_{\infty,s}^l} := \sup_{B \in \mathcal{B}} \inf_{P \in \mathcal{P}_s} |B|^{-l} \operatorname{ess\,sup}_{x \in B} |g(x) - P(x)| < \infty. \qquad (q = \infty)$$

Here $\mathcal{P}_s$ denotes the space of all polynomials (in $n$ variables) of degree at most $s$. We identify two elements of $C_{q,s}^l$ if they are equal almost everywhere.

One may easily verify that: $\|\cdot\|_{C_{q,s}^l}$ is a seminorm, $\mathcal{P}_s \subset C_{q,s}^l$, and $\|g\|_{C_{q,s}^l} = 0 \iff g \in \mathcal{P}_s$. Therefore, $C_{q,s}^l/\mathcal{P}_s$ is a normed linear space. Moreover, the standard arguments show that this space is also complete; hence $C_{q,s}^l/\mathcal{P}_s$ is a Banach space.

The conditions (8.1) and (8.2) can be also written in terms of a quasi-norm $\rho$ associated to a dilation $A$. For example we can equivalently define $C_{q,s}^l$ as the space of all locally $L^q$ functions $g$ on $\mathbb{R}^n$ so that

$$(8.3) \quad \|g\|_{C_{q,s}^l} := \sup_{\substack{x_0 \in \mathbb{R}^n \\ r > 0}} \inf_{P \in \mathcal{P}_s} r^{-l} \left( r^{-1} \int_{\rho(x - x_0) \leq r} |g(x) - P(x)|^q dx \right)^{1/q} < \infty.$$

Indeed, this exactly the case when $\rho$ is the step homogeneous quasi-norm; the general case follows from Lemma 2.4.

The main goal of this section is to prove that the dual of anisotropic Hardy space $H_A^p$ is isomorphic to the Campanato space $C_{q,s}^{1/p-1}/\mathcal{P}_s$, where $(p, q, s)$ is an admissible triplet in the sense of Definition 4.1, see Theorem 8.3. As a consequence, this shows that Campanato spaces $C_{q,s}^l$ depend effectively only the choice of $l$ and not on $q$ or $s$, see Corollary 8.6. Analogous results for Hardy spaces on homogeneous groups were obtained by Folland and Stein [FoS, Chapter 5]. However, careful examination of the arguments in [FoS] reveals a gap in the first part of the proof of [FoS, Theorem 5.3]. The problem is with [FoS, Lemma 5.1] which does not hold unless we assume that a functional $L$ is bounded. Hence, an additional argument is needed. This gap can be filled in the setting of anisotropic Hardy spaces by applying a rather subtle approximation argument inspired by [GR, Chapter III.5].

We start with a basic lemma.

LEMMA 8.2. *Suppose $L$ is a continuous linear functional on $H^p = H^p_{q,s}$, and $(p,q,s)$ is admissible. Then*

$$(8.4) \qquad \|L\|_{(H^p_{q,s})^*} := \sup\{|Lf| : \|f\|_{H^p_{q,s}} \leq 1\} = \sup\{|La| : a \text{ is } (p,q,s)\text{-atom}\}.$$

PROOF. Since every $(p,q,s)$-atom $a$ satisfies $\|a\|_{H^p_{q,s}} \leq 1$ it suffices to show

$$(8.5) \qquad \sup\{|Lf| : \|f\|_{H^p_{q,s}} \leq 1\} \leq \sup\{|La| : a \text{ is } (p,q,s)\text{-atom}\}.$$

Take any $f \in H^p$ with $\|f\|_{H^p_{q,s}} \leq 1$. For every $\varepsilon > 0$ there is an atomic decomposition $f = \sum_i \kappa_i a_i$ with $\sum_i |\kappa_i|^p \leq 1 + \varepsilon$. Therefore

$$|Lf| \leq \sum_i |\kappa_i| |La_i| \leq \sup\{|La| : a \text{ is } (p,q,s)\text{-atom}\} \left(\sum_i |\kappa_i|^p\right)^{1/p}$$

$$\leq (1+\varepsilon)^{1/p} \sup\{|La| : a \text{ is } (p,q,s)\text{-atom}\}.$$

Since $\varepsilon > 0$ is arbitrary we have (8.5). □

Clearly, $\|\cdot\|_{(H^p_{q,s})^*}$ is a norm on $(H^p)^*$ (= space of continuous functionals on $H^p$) which makes $(H^p)^*$ into a Banach space.

THEOREM 8.3. *Suppose $(p,q,s)$ is admissible. Then*

$$(8.6) \qquad (H^p)^* = (H^p_{q,s})^* \equiv C^l_{q',s}/\mathcal{P}_s, \qquad \text{where } 1/q + 1/q' = 1, \ l = 1/p - 1.$$

*If $g \in C^l_{q',s}$ and $f$ is a finite linear combination of $(p,q,s)$-atoms, let $L_g f = \int gf$. Then $L_g$ extends continuously to $H^p$, and every $L \in (H^p)^*$ is of this form. Moreover,*

$$(8.7) \qquad \|g\|_{C^l_{q',s}} = \|L_g\|_{(H^p_{q,s})^*} \qquad \text{for all } g \in C^l_{q',s}.$$

PROOF. For any $B \in \mathcal{B}$ define $\pi_B : L^1(B) \to \mathcal{P}_s$ the natural projection defined using the Riesz Lemma

$$(8.8) \qquad \int_B (\pi_B f(x)) Q(x) dx = \int_B f(x) Q(x) dx \qquad \text{for all } f \in L^1(B), Q \in \mathcal{P}_s.$$

We claim that there is a universal constant (depending only on $s$) such that

$$(8.9) \qquad \sup_{x \in B} |\pi_B f(x)| \leq C \frac{1}{|B|} \int_B |f(x)| dx.$$

Indeed, if $B = B_0$ then we find an orthonormal basis $\{Q_\alpha : |\alpha| \leq s\}$ of $\mathcal{P}_s$ with respect to the $L^2(B_0)$ norm. Since

$$\pi_{B_0} f = \sum_{|\alpha| \leq s} \left(\int_{B_0} f(x) \overline{Q_\alpha(x)} dx\right) Q_\alpha.$$

the claim follows for $B = B_0$. Since $\pi_{B_k} f = (D_{A^{-k}} \circ \pi_{B_0} \circ D_{A^k})f$ and $\pi_{y+B_k} f = (\tau_y \circ \pi_{B_k} \circ \tau_{-y})f$ the claim (8.9) follows for arbitrary $B = y + B_k$.

For $1 \le q \le \infty$ and $B \in \mathcal{B}$ we define the closed subspace $L_0^q(B) \subset L^q(B)$ by $L_0^q(B) = \{f \in L^q(B) : \pi_B f = 0\}$. We will identify $L^q(B)$ with the subspace of $L^q(\mathbb{R}^n)$ consisting of functions vanishing outside $B$. With this identification, if $f \in L_0^q(B)$ then $|B|^{1/q-1/p} \|f\|_q^{-1} f$ is a $(p,q,s)$-atom.

Suppose $L \in (H^p)^* = (H_{q,s}^p)^*$. By Lemma 8.2

$$(8.10) \qquad |Lf| \le \|L\|_{(H_{q,s}^p)^*} |B|^{1/p-1/q} \|f\|_q, \qquad \text{for } f \in L_0^q(B).$$

Therefore, $L$ provides a bounded linear functional on $L_0^q(B)$ which can be extended by the Hahn-Banach Theorem to the whole space $L^q(B)$ without increasing its norm. Suppose, momentarily, that $q < \infty$. By the duality $L^q(B)^* = L^{q'}(B)$, where $1/q + 1/q' = 1$, there exists $h \in L^{q'}(B)$ such that $Lf = \int fh$ for all $f \in L_0^q(B)$. In particular, $L^\infty(B) \subset L^q(B)$ implies there is $h \in L^{q'}(B) \subset L^1(B)$ such that $Lf = \int fh$ for all $f \in L_0^\infty(B)$. Therefore, also for $q = \infty$, there exists $h \in L^{q'}(B)$ such that $Lf = \int fh$ for all $f \in L_0^q(B)$. If $h'$ is another element of $L^{q'}(B)$ such that $Lf = \int fh'$ for all $f \in L_0^q(B)$ then $h - h' \in \mathcal{P}_s$. We can say even more. Suppose that for some $h, h' \in L^1(B)$ we have $Lf = \int_B fh = \int_B fh'$ for all $f \in L_0^\infty(B)$ then $h - h' \in \mathcal{P}_s$. Indeed, for all $f \in L^\infty(B)$

$$0 = \int_B (f - \pi_B f)(h - h') = \int_B f(h - h') - \int_B (\pi_B f)(\pi_B(h - h'))$$
$$= \int_B f(h - h') - \int_B f(\pi_B(h - h')) = \int_B f((h - h') - \pi_B(h - h')),$$

hence $(h - h')(x) = \pi_B(h - h')(x)$ for a.e. $x \in B$. Therefore, after changing values of $h$ (or $h'$) on a set of measure zero we have $h - h' \in \mathcal{P}_s$. Hence, the function $h$ is unique up to a polynomial of degree at most $s$ regardless of the exponent $1 \le q \le \infty$. Therefore, $h \in \bigcap_{q<\infty} L^q(B)$ for $p = 1$, and $h \in L^\infty(B)$ for $0 < p < 1$.

For $k = 1, 2, \ldots$ let $g_k$ be the unique element of $L^{q'}(B_k)$ such that $Lf = \int fg_k$ for all $f \in L_0^q(B_k)$ and $\pi_{B_0} g_k = 0$. The preceding arguments show that $g_k|_{B_j} = g_j$ for $j < k$. Therefore, we can define a locally $L^{q'}$ function $g$ on $\mathbb{R}^n$ by setting $g(x) = g_k(x)$ for $x \in B_k$. If $f$ is a finite linear combination of $(p,q,s)$-atoms then $Lf = \int fg$.

By (8.10), for any $B \in \mathcal{B}$ the norm of $g$ as a linear functional on $L_0^q(B)$ satisfies

$$(8.11) \qquad \|g\|_{L_0^q(B)^*} \le \|L\|_{(H_{q,s}^p)^*} |B|^{1/p-1/q}.$$

We claim that

$$(8.12) \qquad \|g\|_{L_0^q(B)^*} = \inf_{P \in \mathcal{P}_s} \|g - P\|_{L^{q'}(B)}.$$

This is, at least for $q < \infty$, an immediate consequence of an elementary fact from functional analysis

$$L_0^q(B)^* = L^q(B)^* / L_0^q(B)^\perp = L^{q'}(B)/\mathcal{P}_s,$$

where $L_0^q(B)^\perp = \{h \in L^q(B)^* : h|_{L_0^q(B)} = 0\}$ denotes the annihilator of the subspace $L_0^q(B) \subset L^q(B)$. The special case of (8.12) for $q = \infty$ requires similar

## 8. DUALS OF $H^p$

arguments and the fact that $g \in L^1(B)$ and $||g-P||_{L^\infty(B)^*} = ||g-P||_{L^1(B)}$ for any $P \in \mathcal{P}_s$. Combining (8.11) and (8.12) we have for $l = 1/p - 1$

$$(8.13) \quad ||g||_{C^l_{q',s}} = \sup_{B \in \mathcal{B}} |B|^{-l-1/q'} ||g||_{L^q_0(B)^*} \leq ||L||_{(H^p_{q,s})^*},$$

which completes the first part of the proof of Theorem 8.3.

Conversely, suppose $g \in C^l_{q',s}$. Our goal is to show that functional $L_g f = \int gf$ defined initially for $f \in \Theta^q_s$, where

$$(8.14) \quad \Theta^q_s = \{f \in L^q(\mathbb{R}^n) : \operatorname{supp} f \text{ is compact and } \int_{\mathbb{R}^n} f(x) x^\alpha dx = 0 \text{ for } |\alpha| \leq s\},$$

extends to a bounded functional on $H^p_{q,s}$ and $||L_g||_{(H^p_{q,s})^*} \leq ||g||_{C^l_{q',s}}$.

Suppose $a$ is $(p,q,s)$-atom associated to the dilated ball $B = x_0 + B_k$. Since $\int ga = \int (g-P)a$ for all $P \in \mathcal{P}_s$ we have

$$(8.15) \quad \begin{aligned} |L_g a| &= \left| \int ga \right| = \inf_{P \in \mathcal{P}_s} \left| \int (g-P)a \right| \\ &\leq \left( \int_B |a|^q \right)^{1/q} \left( \inf_{P \in \mathcal{P}_s} \int |g-P|^{q'} \right)^{1/q'} \\ &\leq |B|^{1/q-1/p} |B|^{l+1/q'} \left( \inf_{P \in \mathcal{P}_s} \int |g-P|^{q'} \right)^{1/q'} = ||g||_{C^l_{q',s}}. \end{aligned}$$

At this moment the reader might be tempted to use Lemma 8.2. That is we can try to define $L_g f$ for arbitrary $f \in H^p$ by using the atomic decomposition of $f$,

$$(8.16) \quad L_g f = \sum_{i=1}^\infty \kappa_i L_g a_i, \qquad \text{where } f = \sum_{i=1}^\infty \kappa_i a_i.$$

Since for every $\varepsilon > 0$ we can choose a decomposition so that $\sum_{i=1}^\infty |\kappa_i|^p \leq (1+\varepsilon) ||f||^p_{H^p_{q,s}}$ we have

$$(8.17) \quad |L_g f| \leq \sum_{i=1}^\infty |\kappa_i| |L_g a_i| \leq ||g||_{C^l_{q',s}} \left( \sum_{i=1}^\infty |\kappa_i|^p \right)^{1/p} \leq (1+\varepsilon) ||g||_{C^l_{q',s}} ||f||_{H^p_{q,s}}.$$

This may seem to show that $L_g$ is bounded with the appropriate constant.

The problem with this argument is the issue of well-definedness of $L_g$. Namely, given $g \in C^l_{q',s}$, we ought to make sure that if $f \in \Theta^q_s$ has an atomic decomposition $f = \sum_{i=1}^\infty \kappa_i a_i$ then necessarily

$$\int fg = \sum_{i=1}^\infty \kappa_i \int a_i g.$$

As we will see this fact is very non-trivial and it requires a rather subtle approximation argument. For the classical isotropic case, the interested reader is directed to [GR, Section III.5], where detailed exposition of this fact can be found. Note that there is no problem if $g$ belongs to the Schwartz class, since the atomic decomposition of $f$ converges in the sense of distributions.

The following lemma comes to the rescue.

LEMMA 8.4. *Suppose that $(p,q,s)$ is admissible and $f \in \Theta_s^q$, where $\Theta_s^q$ is given by (8.14). Suppose $g \in C_{q',s}^l$, $1/q + 1/q' = 1$, $l = 1/p - 1$. There exists $\tilde{s} \geq s$ so that if $f$ is decomposed into $f = \sum_{i=1}^\infty \kappa_i a_i$, where $\sum_{i=1}^\infty |\kappa_i|^p < \infty$ and $a_i$'s are $(p,q,\tilde{s})$-atoms, then*

$$(8.18) \qquad L_g f = \int fg = \sum_{i=1}^\infty \kappa_i \int a_i g = \sum_{i=1}^\infty \kappa_i L_g a_i.$$

Given Lemma 8.4 the rest of the proof is immediate. Suppose $f \in \Theta_s^q$. By Theorem 6.5 we can find an atomic decomposition of $f = \sum_{i=1}^\infty \kappa_i a_i$, where

$$\left(\sum_{i=1}^\infty |\kappa_i|^p\right)^{1/p} \leq 2\|f\|_{H_{q,\tilde{s}}^p} \leq C\|f\|_{H_{q,s}^p},$$

and $a_i$'s are $(p,q,\tilde{s})$-atoms. By (8.15) and (8.18)

$$(8.19) \qquad |L_g f| \leq \sum_{i=1}^\infty |\kappa_i| |L_g a_i| \leq \|g\|_{C_{q',s}^l} \left(\sum_{i=1}^\infty |\kappa_i|^p\right)^{1/p} \leq C\|g\|_{C_{q',s}^l} \|f\|_{H_{q,s}^p}.$$

Therefore, $L_g$ extends uniquely to a bounded functional on $H_{q,s}^p$. Furthermore, by Lemma 8.2 and (8.15) $\|L_g\|_{(H_{q,s}^p)^*} \leq \|g\|_{C_{q',s}^l}$. This finishes the proof of Theorem 8.3. □

REMARK. As a consequence of Theorem 8.3 we conclude that if we write $f = \sum_{i=1}^\infty \kappa_i a_i \in \Theta_s^q$, where $\sum_{i=1}^\infty |\kappa_i|^p < \infty$ and $a_i$'s are $(p,q,s)$-atoms then (8.18) still holds. As an indication why this result is non-trivial we recall the following observation due to Meyer, see [GR, MTW].

In Definition 4.3 of the atomic norm $\|f\|_{H_{q,s}^p}$, it is not legitimate to take the infimum only over finite linear combinations of atoms even in the case when $f$ admits such a finite decomposition, e.g., when $f$ is itself an atom. Indeed, such infimum may be much larger than $\|f\|_{H_{q,s}^p}$, which is evidenced by an example due to Meyer, see [GR, Chapter III.8.3]. Therefore, even though finite linear combinations of atoms are dense in $H^p$, in order to compute their $H^p$ norm, it is not enough to look only at finite decompositions.

To show Lemma 8.4 we need the following approximation lemma.

LEMMA 8.5. *Suppose $g \in C_{q',s}^l$, where $l \geq 0$, $1 \leq q' \leq \infty$, $s = 0, 1, \ldots$ There exists $\tilde{s} \geq s$, a constant $C > 0$ independent of $g$, and a sequence of test functions $(g_k)_{k \in \mathbb{N}} \subset \mathcal{S}$ so that*

$$(8.20) \qquad \|g_k\|_{C_{q',\tilde{s}}^l} \leq C\|g\|_{C_{q',s}^l} \qquad \text{for all } k \in \mathbb{N},$$

*and $(1/q + 1/q' = 1)$,*

$$(8.21) \qquad \lim_{k \to \infty} \int_{\mathbb{R}^n} f(x) g_k(x) dx = \int_{\mathbb{R}^n} f(x) g(x) dx \qquad \text{for all } f \in \Theta_s^q, .$$

We are going to use two simple observations about Campanato spaces. Firstly, note that if $D_A g(x) = g(Ax)$ then

$$\|D_A g\|_{C^l_{q',s}} = \sup_{\substack{x_0 \in \mathbb{R}^n \\ k \in \mathbb{Z}}} \inf_{P \in \mathcal{P}_s} |B_k|^{-l} \left( \frac{1}{|B_k|} \int_{x_0 + B_k} |g(Ax) - P(x)|^{q'} dx \right)^{1/q'}$$

(8.22)
$$= \sup_{\substack{x_0 \in \mathbb{R}^n \\ k \in \mathbb{Z}}} \inf_{P \in \mathcal{P}_s} b^l |B_{k+1}|^{-l} \left( \frac{1}{|B_{k+1}|} \int_{Ax_0 + B_{k+1}} |g(x) - P(x)|^{q'} dx \right)^{1/q'}$$

$$= b^l \|g\|_{C^l_{q',s}}.$$

Secondly, we can define an equivalent norm on $C^l_{q',s}$ by setting

(8.23) $\quad \||g\||_{C^l_{q',s}} = \sup_{B \in \mathcal{B}} |B|^{-l} \left( \frac{1}{|B|} \int_B |g(x) - \pi_B g(x)|^{q'} dx \right)^{1/q'} \quad (1 \leq q' < \infty),$

(8.24) $\quad \||g\||_{C^l_{\infty,s}} = \sup_{B \in \mathcal{B}} |B|^{-l} \operatorname{ess\,sup}_{x \in B} |g(x) - \pi_B g(x)| \quad (q' = \infty).$

where $\pi_B g$ is the natural projection on $\mathcal{P}_s$ given by (8.8). Indeed, for any $B \in \mathcal{B}$ and $P \in \mathcal{P}_s$ by the Minkowski inequality we have

$$\left( \frac{1}{|B|} \int_B |g(x) - \pi_B g(x)|^{q'} dx \right)^{1/q'}$$

$$\leq \left( \frac{1}{|B|} \int_B |g(x) - P(x)|^{q'} dx \right)^{1/q'} + \left( \frac{1}{|B|} \int_B |P(x) - \pi_B g(x)|^{q'} dx \right)^{1/q'}$$

$$\leq \left( \frac{1}{|B|} \int_B |g(x) - P(x)|^{q'} dx \right)^{1/q'} + C \frac{1}{|B|} \int_B |g(x) - P(x)| dx$$

$$\leq (C+1) \left( \frac{1}{|B|} \int_B |g(x) - P(x)|^{q'} dx \right)^{1/q'}.$$

using $P(x) - \pi_B g(x) = \pi_B(P - g)(x)$ and (8.9). Therefore we have

(8.25) $\quad \|g\|_{C^l_{q',s}} \leq \||g\||_{C^l_{q',s}} \leq (C+1)\|g\|_{C^l_{q',s}} \quad \text{for all } g \in C^l_{q',s}.$

PROOF OF LEMMA 8.5. Suppose $g \in C^l_{q',s}$. Take a nonnegative function $\varphi \in C^\infty$ with compact support and $\int \varphi = 1$. Let $\varphi_k(x) = b^{-k} \varphi(A^{-k} x)$. If $q' < \infty$ then

$$g * \varphi_k \to g \quad \text{in } L^{q'}_{\text{loc}}(\mathbb{R}^n) \quad \text{as } k \to -\infty.$$

Indeed, it suffices to apply Theorem 6.8 in $L^{q'}$ for truncations $g \mathbf{1}_{\mathbf{B}(0,r)}$ for sufficiently large $r > 0$. Therefore

(8.26) $\quad \int_{\mathbb{R}^n} f(x)(g * \varphi_k)(x) dx \to \int_{\mathbb{R}^n} f(x) g(x) dx \quad \text{as } k \to \infty \quad \text{for all } f \in \Theta^q_s.$

If $q' = \infty$ we use Theorem 3.7 applied for truncations $g \mathbf{1}_{\mathbf{B}(0,r)}$ for sufficiently large $r > 0$ to obtain

$$(g * \varphi_k)(x) \to g(x) \quad \text{as } k \to -\infty \quad \text{for a.e. } x \in \mathbb{R}^n.$$

By the Lebesgue Dominated Convergence Theorem (8.26) holds also for $q' = \infty$. Clearly, $g * \varphi_k \in C^\infty$ and moreover

(8.27) $$\|g * \varphi_k\|_{C^l_{q',s}} \leq \||g\|\|_{C^l_{q',s}} \qquad \text{for all } k \in \mathbb{N}.$$

Indeed, for every $B \in \mathcal{B}$ and $k \in \mathbb{N}$ define a function $P_k$ by

$$P_k(x) = \int_{\mathbb{R}^n} \pi_{-y+B} g(x-y) \varphi_k(y) dy.$$

Since we can write $\pi_{-y+B} g(x-y) = \sum_{|\alpha| \leq s} c_\alpha(y)(x-y)^\alpha$ and the coefficients $c_\alpha(y)$ are continuous functions of $y$, $P_k$ is a polynomial of degree $\leq s$. By the Minkowski inequality

$$\left(\frac{1}{|B|} \int_B |(g * \varphi_k)(x) - P_k(x)|^{q'} dx\right)^{1/q'}$$

$$= \left(\frac{1}{|B|} \int_B \left|\int_{\mathbb{R}^n} (g(x-y) - \pi_{-y+B} g(x-y)) \varphi_k(y) dy\right|^{q'} dx\right)^{1/q'}$$

$$\leq \int_{\mathbb{R}^n} \left(\frac{1}{|B|} \int_B |g(x-y) - \pi_{-y+B} g(x-y)|^{q'} dx\right)^{1/q'} |\varphi_k(y)| dy$$

$$= \int_{\mathbb{R}^n} \left(\frac{1}{|-y+B|} \int_{-y+B} |g(z) - \pi_{-y+B} g(z)|^{q'} dz\right)^{1/q'} \varphi_k(y) dy \leq \||g\|\|_{C^l_{q',s}} |B|^l.$$

This shows (8.27).

Formulae (8.26) and (8.27) suggest that it is enough to show the lemma for $g \in C^l_{q',s} \cap C^\infty$. That is indeed the case. Let $\phi \in C^\infty$ be such that $\operatorname{supp} \phi \subset B_0$, $0 \leq \phi(x) \leq 1$, and $\phi(x) = 1$ for $x \in B_{-1}$. We claim that there is a constant $C$ and $\tilde{s} \geq s$ such that

(8.28) $$\|(g - \pi_{B_0} g)\phi\|_{C^l_{q',\tilde{s}}} \leq C \|g\|_{C^l_{q',s}} \qquad \text{for all } g \in C^l_{q',s} \cap C^\infty.$$

Indeed, take any $g \in C^l_{q',s} \cap C^\infty$ with $\||g\|\|_{C^l_{q',s}} \leq 1$. For brevity, we only consider the case $q' < \infty$; the case $q' = \infty$ uses a similar argument. Let $G = g - \pi_{B_0} g$. Since $\operatorname{supp} \phi \subset B_0$, $\int_{\mathbb{R}^n} |G\phi|^{q'} \leq \int_{B_0} |G|^{q'} \leq |B_0|^{lq'+1} = 1$. Therefore, if we take ball $B \in \mathcal{B}$, $B = x_0 + B_k$ and $k \geq 0$ then

$$|B|^{-l} \left(\frac{1}{|B|} \int_B |G(x)\phi(x)|^{q'} dx\right)^{1/q'} \leq 1.$$

Hence, to show (8.28) we can restrict to balls $B = x_0 + B_k$ with $k < 0$. Let $P_1 = \pi_B g = \pi_B G$. By (8.9)

(8.29) $$\left(\frac{1}{|B|} \int_B |P_1(x)|^{q'} dx\right)^{1/q'} \leq C \left(\frac{1}{|B|} \int_B |G(x)|^{q'} dx\right)^{1/q'} \leq C b^{-k/q'}.$$

Let $P_2(x)$ be the Taylor polynomial of $\phi$ at $x_0$ of degree $r$ (to be specified later), i.e., $P_2(x) = \sum_{|\alpha| \leq r} \partial^\alpha \phi(x_0)(x-x_0)^\alpha/\alpha!$. By the Taylor Theorem the remainder satisfies $|\phi(x) - P_2(x)| \leq C'|x-x_0|^{r+1}$ with the constant $C'$ independent of $x_0$.

## 8. DUALS OF $H^p$

Finally, let $P(x) = P_1(x)P_2(x)$ be a polynomial of degree at most $\tilde{s} = s + r$. By the Minkowski inequality and (8.29)

$$\left(\int_B |G(x)\phi(x) - P(x)|^{q'} dx\right)^{1/q'} \leq$$

$$\left(\int_B |G(x)\phi(x) - P_1(x)\phi(x)|^{q'} dx\right)^{1/q'} + \left(\int_B |P_1(x)\phi(x) - P_1(x)P_2(x)|^{q'} dx\right)^{1/q'}$$

$$\leq \|\phi\|_\infty \left(\int_B |G(x) - P_1(x)|^{q'} dx\right)^{1/q'} + \sup_{x \in B} |\phi(x) - P_2(x)| \left(\int_B |P_1(x)|^{q'} dx\right)^{1/q'}$$

$$\leq |B|^{l+1/q'} + b^{-k/q'} C' \sup_{x \in x_0 + B_k} |x - x_0|^{r+1} \leq |B|^{l+1/q'} + b^{-k/q'} C'(c\lambda_-^k)^{r+1}$$

$$\leq |B|^{l+1/q'} + Cb^{-k/q'} b^{k(l+2/q')} = (C+1)|B|^{l+1/q'}.$$

Here we need to choose $r$ large enough such that $\lambda_-^{r+1} \geq b^{l+2/q'}$, i.e., take $r = \lfloor (l + 2/q') \ln b / \ln \lambda_- \rfloor$. Since $B \in \mathcal{B}$ is arbitrary this shows (8.28) with $\tilde{s} = s + r$.

We are now ready to define a sequence $(g_k)_{k \in \mathbb{N}} \subset \mathcal{S}$ for $g \in C^l_{q',s} \cap C^\infty$. Let $\tilde{g}_k = D_{A^k} g$ and $g_k = D_{A^{-k}}((\tilde{g}_k - \pi_{B_0} \tilde{g}_k)\phi)$. By (8.22) and (8.28)

$$\|(\tilde{g}_k - \pi_{B_0} \tilde{g}_k)\phi\|_{C^l_{q',\tilde{s}}} \leq C \|\tilde{g}_k\|_{C^l_{q',s}} = C b^{kl} \|g\|_{C^l_{q',s}}$$

Therefore (8.20) holds, since

$$(8.30) \qquad \|g_k\|_{C^l_{q',\tilde{s}}} = b^{-kl} \|(\tilde{g}_k - \pi_{B_0} \tilde{g}_k)\phi\|_{C^l_{q',\tilde{s}}} \leq C \|g\|_{C^l_{q',s}}.$$

Moreover,

$$(8.31) \quad g_k(x) = g(x) - (D_{A^{-k}} \circ \pi_{B_0} \circ D_{A^k}) g(x) = g(x) - \pi_{B_k} g(x) \qquad \text{for } x \in B_{k-1}.$$

Thus (8.21) also holds.

To end the proof we must relax the assumption that $g \in C^\infty$. Suppose $g \in C^l_{q',s}$ is arbitrary. Define the sequence $(g_k)_{k \in \mathbb{N}} \subset \mathcal{S}$ by $g_k = D_{A^{-k}}((\tilde{g}_k - \pi_{B_0} \tilde{g}_k)\phi)$, where $\tilde{g}_k = D_{A^k}(g * \varphi_k)$. Combining (8.27) and (8.30) yields (8.20), whereas (8.26) and (8.31) yield (8.21), completing the proof of Lemma 8.5. □

To completely finish the proof of the duality of $H^p$ spaces we must establish Lemma 8.4.

PROOF OF LEMMA 8.4. Suppose $f \in \Theta^q_s$ is decomposed into $f = \sum_{i=1}^\infty \kappa_i a_i$, where $\sum_{i=1}^\infty |\kappa_i|^p < \infty$ and $a_i$'s are $(p,q,\tilde{s})$-atoms, where $\tilde{s} \geq s$ is the same as in Lemma 8.5. Suppose also that $g \in C^l_{q',s}$, $1/q + 1/q' = 1$, $l = 1/p - 1$. Let $(g_k)_{k \in \mathbb{N}} \subset \mathcal{S}$ be a sequence guaranteed by Lemma 8.5. For every $k \in \mathbb{N}$ we have

$$(8.32) \qquad L_{g_k} f = \int f g_k = \sum_{i=1}^\infty \kappa_i \int a_i g_k = \sum_{i=1}^\infty \kappa_i L_{g_k} a_i,$$

since convergence in $H^p$ implies convergence in $\mathcal{S}'$ by Theorem 4.5. By (8.21)

$$\lim_{k \to \infty} \int_{\mathbb{R}^n} a_i(x) g_k(x) dx = \int_{\mathbb{R}^n} a_i(x) g(x) dx \qquad \text{for all } i \in \mathbb{N}.$$

By (8.15) and (8.20) we have $|L_{g_k} a_i| \leq \|g_k\|_{C^l_{q',\tilde{s}}} \leq C\|g\|_{C^l_{q',s}}$. Since $\sum_{i=1}^{\infty} |\kappa_i| \leq (\sum_{i=1}^{\infty} |\kappa_i|^p)^{1/p} < \infty$ we can take limit as $k \to \infty$ in (8.32) by the Lebesgue Dominated Convergence Theorem applied to the counting measure on $\mathbb{N}$. This shows (8.18). □

As an immediate consequence of Theorems 6.5 and 8.3 we have

COROLLARY 8.6. *Suppose that $l \geq 0$, $1 \leq q, q' \leq \infty$ ($q, q' < \infty$ if $l = 0$), and $s, s' \geq \lfloor l \ln b / \ln \lambda_- \rfloor$. Then $C^l_{q,s} = C^l_{q',s'}$ and the seminorms $\|\cdot\|_{C^l_{q,s}}$, $\|\cdot\|_{C^l_{q',s'}}$ are equivalent. If $l = 0$ then $C^0_{q,s} = C^0_{1,0}$ is the space of BMO (bounded mean oscillation).*

The following simple proposition is very useful.

PROPOSITION 8.7. *Suppose $g \in C^N(\mathbb{R}^n)$ is bounded and all partial derivatives $\partial^\alpha g$ of order $N$, $|\alpha| = N$, are bounded. Then for every $0 \leq l \leq N \ln \lambda_- / \ln b$ and $s \geq \lfloor l \ln b / \ln \lambda_- \rfloor$, $g \in C^l_{\infty,s}$.*

PROOF. The case $N = 0$ is trivial. If $N \geq 1$ then by Corollary 8.6 it suffices to show that $g \in C^l_{\infty, N-1}$. We need to show that there exists a constant $C$ such that for every dilated ball $B \in \mathcal{B}$,

$$(8.33) \qquad \inf_{P \in \mathcal{P}_{N-1}} \sup_{x \in B} |g(x) - P(x)| \leq C|B|^l.$$

Since $g$ is bounded, (8.33) trivially holds for large balls with $|B| \geq 1$. Take any $B = x_0 + B_k \in \mathcal{B}$, where $x_0 \in \mathbb{R}^n$ and $k < 0$. Let $P(x) \in \mathcal{P}_{N-1}$ be the Taylor polynomial of $g$ at the point $x_0$ of order $N - 1$. By the Taylor Remainder Theorem

$$|g(x) - P(x)| \leq C \sup_{|\alpha|=N} \|\partial^\alpha g\|_\infty \sup_{z \in B_k} |z|^N \leq C(\lambda_-)^{Nk} \qquad \text{for all } x \in B.$$

Since $(\lambda_-)^N \geq b^l$, we obtain (8.33). □

There are redundant alternative ways of defining Campanato spaces. Let $\Delta_h$ denote the difference operator by a vector $h \in \mathbb{R}^n$, i.e., $\Delta_h g(x) = g(x + h) - g(x)$. Suppose that $l > 0$ and $s \geq 1$. Define by $C^l_s$ the space of all continuous functions $g$ on $\mathbb{R}^n$ such that

$$|\Delta_{h_1} \ldots \Delta_{h_s} g(x)| \leq c(\rho(h_1) + \ldots + \rho(h_s))^l \qquad \text{for all } x \in \mathbb{R}^n, (h_1, \ldots, h_s) \in (\mathbb{R}^n)^s$$

for some constant $c$. The infimum over all constants $c$ satisfying the above defines the seminorm of $g$ in $C^l_s$. Some effort is needed to show that $C^l_{q,s} = C^l_s$ for sufficiently large $s$ and $1 \leq q \leq \infty$, see [Gr, Ja, JTW, Kr] for the isotropic case.

EXAMPLE. The best known example of Campanato spaces occurs when the dilation $A$ defines the usual isotropic structure on $\mathbb{R}^n$, for example if $A = 2Id$. In this case the Campanato spaces $C^l_{q,s}(\mathbb{R}^n)$ coincide with the *homogeneous Hölder spaces* $\dot{C}^\gamma(\mathbb{R}^n)$ (sometimes also called *Lipshitz spaces* $\Lambda_\gamma(\mathbb{R}^n)$) with $\gamma = nl$, for the proof see [GR, Section III.5].

The space $\dot{C}^\gamma(\mathbb{R}^n)$ for $\gamma > 0$ is defined as follows. Let $\gamma = \lceil \gamma \rceil + \gamma' - 1$, $0 < \gamma' \leq 1$. The space $\dot{C}^\gamma(\mathbb{R}^n)$ consists of all functions $g$ in $C^{\lceil \gamma \rceil - 1}(\mathbb{R}^n)$ ($\lceil \gamma \rceil - 1$

Then $f \in H^p$ with $||f||_{H^p_{q,s}} \leq C'$, where $C'$ depends only on $C$.

We remark that (9.9) is meaningful since $|f(x)|(1+|x|^s)$ is integrable by (9.7) and (9.8). Indeed, by Lemma 3.2

$$|f(x)||x|^s \leq C\rho(A^{-k}x)^{-\delta}c'\rho(x)^{s\ln\lambda_+/\ln b} = Cc'b^{k\delta}\rho(x)^{s\ln\lambda_+/\ln b-\delta} \quad \text{for } \rho(x) \geq 1.$$

PROOF. Given a ball $B \in \mathcal{B}$ consider the natural projection $\pi_B : L^1(B) \to \mathcal{P}_s$ given by (8.8). Define the complementary projection $\tilde{\pi}_B = Id - \pi_B$, i.e., $\tilde{\pi}_B f = f - \pi_B f$. By (8.9) $\tilde{\pi}_B$ is bounded on $L^q(B)$, i.e.,

$$(9.10) \qquad ||\tilde{\pi}_B f||_{L^q(B)} \leq C_0 ||f||_{L^q(B)},$$

with the constant $C_0$ independent of $B \in \mathcal{B}$. Moreover,

$$\int_B \tilde{\pi}_B f(x) x^\alpha = 0 \qquad \text{for } |\alpha| \leq s.$$

We want to represent $f$ as a combination of atoms. To do this define the sequence of functions $(g_j)_{j=k}^\infty$ by $g_j = \tilde{\pi}_{B_j}(f)\mathbf{1}_{B_j}$. Clearly, supp $g_j \subset B_j$ for $j \geq k$. Since $||g_k||_q \leq C_0 ||f\mathbf{1}_{B_k}||_q \leq C_0 C |B_k|^{1/q-1/p}$ and $g_k$ has vanishing moments up to order $s$, $g_k$ is at most $C_0 C$ multiplicity of some $(p,q,s)$-atom (namely $(C_0 C)^{-1} g_k$).

We claim that $g_j \to f$ in $L^1$ (and hence in $\mathcal{S}'$) as $j \to \infty$. It suffices to show that $||\pi_{B_j} f||_{L^1(B_j)} \to 0$ as $j \to \infty$. Indeed, let $\{Q_\alpha : |\alpha| \leq s\}$ be an orthonormal basis of $\mathcal{P}_s$ with respect to the $L^2(B_0)$ norm. By the argument used to show (8.9) we have

$$(9.11) \quad \begin{aligned} \pi_{B_j} f &= (D_{A^{-j}} \circ \pi_{B_0} \circ D_{A^j}) f = \sum_{|\alpha| \leq s} \left( \int_{B_0} D_{A^j} f(x) \overline{Q_\alpha(x)} dx \right) D_{A^{-j}} Q_\alpha \\ &= \sum_{|\alpha| \leq s} \left( \int_{B_j} f(x) \overline{Q_\alpha(A^{-j}x)} dx \right) b^{-j} D_{A^{-j}} Q_\alpha, \end{aligned}$$

where $D_A f(x) = f(Ax)$. We also have $||D_{A^{-j}} Q_\alpha||_{L^1(B_j)} = b^j ||Q_\alpha||_{L^1(B_0)} \leq b^j$ and

$$\int_{B_j} f(x) \overline{Q_\alpha(A^{-j}x)} dx = -\int_{B_j^c} f(x) \overline{Q_\alpha(A^{-j}x)} dx \to 0 \quad \text{as } j \to \infty,$$

by (9.9) and the uniform boundedness of coefficients of the polynomials $Q_\alpha(A^{-j}x)$ for $j \geq 0$. This shows

$$(9.12) \qquad f = g_k + \sum_{j=k}^\infty (g_{j+1} - g_j) \qquad \text{with convergence in } L^1(\mathbb{R}^n).$$

In fact, we will prove that we also have convergence in $H^p$ by showing that $g_{j+1} - g_j$ are appropriate multiples of $(p, \infty, s)$-atoms supported on $B_{j+1}$. Indeed,

$$(9.13) \quad \begin{aligned} ||g_{j+1} - g_j||_\infty &= ||\tilde{\pi}_{B_{j+1}}(f)\mathbf{1}_{B_{j+1}} - \tilde{\pi}_{B_j}(f)\mathbf{1}_{B_j}||_\infty \\ &= ||f\mathbf{1}_{B_{j+1}\setminus B_j} - \mathbf{1}_{B_{j+1}} \pi_{B_{j+1}} f + \mathbf{1}_{B_j} \pi_{B_j} f||_\infty \\ &\leq ||f\mathbf{1}_{B_{j+1}\setminus B_j}||_\infty + ||\mathbf{1}_{B_{j+1}} \pi_{B_{j+1}} f||_\infty + ||\mathbf{1}_{B_j} \pi_{B_j} f||_\infty. \end{aligned}$$

By (9.8)

$$||f\mathbf{1}_{B_{j+1}\setminus B_j}||_\infty \leq Cb^{-k/p} b^{\delta(k-j)} = Cb^{-j/p} b^{(\delta-1/p)(k-j)}.$$

times continuously differentiable) with the norm

$$||g||_{\dot{C}^\gamma(\mathbb{R}^n)} = \sup_{|\alpha|=\lceil\gamma\rceil-1} \sup_{h\in\mathbb{R}^n\setminus\{0\}} |h|^{-\gamma'} ||\Delta_h \partial^\alpha g||_\infty \qquad \text{if } \gamma' < 1,$$

$$||g||_{\dot{C}^\gamma(\mathbb{R}^n)} = \sup_{|\alpha|=\lceil\gamma\rceil-1} \sup_{h\in\mathbb{R}^n\setminus\{0\}} |h|^{-1} ||\Delta_h^2 \partial^\alpha g||_\infty \qquad \text{if } \gamma' = 1.$$

For example, if $0 < \gamma < 1$ then $\dot{C}^\gamma(\mathbb{R}^n)$ is the usual Hölder space consisting of all functions $g$ satisfying

$$|g(x+h) - g(x)| \leq c|h|^\gamma \qquad \text{for all } x, h \in \mathbb{R}^n,$$

and if $\gamma = 1$ then $\dot{C}^1(\mathbb{R}^n)$ is the *Zygmund class* [Zy] consisting of all functions $g$ satisfying

$$|g(x+2h) - 2g(x+h) + g(x)| \leq c|h| \qquad \text{for all } x, h \in \mathbb{R}^n.$$

## 9. Calderón-Zygmund singular integrals on $H^p$

In this section we present the theory of Calderón-Zygmund singular integrals on the anisotropic $H^p$ spaces for $0 < p \leq 1$. For $p > 1$ this theory reduces to studying $L^p$ spaces and therefore follows from the general theory of Calderón-Zygmund singular integrals on the spaces of homogeneous type, see [St2, Chapter 1.5].

We start with some preliminaries. Let $T : \mathcal{S}(\mathbb{R}^n) \to \mathcal{S}'(\mathbb{R}^n)$ be a continuous linear operator. By the Schwartz Kernel Theorem there exists $S \in \mathcal{S}'(\mathbb{R}^n \times \mathbb{R}^n)$ such that

$$(9.1) \qquad \langle T(f), g \rangle = \langle S, g \otimes f \rangle \qquad \text{for all } f, g \in \mathcal{S}(\mathbb{R}^n).$$

Let $\Omega = \{(x, y) \in \mathbb{R}^n \times \mathbb{R}^n : x \neq y\}$. We say that a distribution $S$ is *regular* on $\Omega$ if there exists a locally integrable function $K(x, y)$ on $\Omega$ such that

$$(9.2) \qquad S(G) = \int_\Omega K(x, y) G(x, y) dx dy \qquad \text{for all } G \in \mathcal{S}(\mathbb{R}^n \times \mathbb{R}^n), \overline{\operatorname{supp} G} \subset \Omega.$$

DEFINITION 9.1. Let $T : \mathcal{S}(\mathbb{R}^n) \to \mathcal{S}'(\mathbb{R}^n)$ be a continuous linear operator. We say that $T$ is a *Calderón-Zygmund operator* (with respect to a dilation $A$ with a quasi-norm $\rho$) if there are constants $C > 0$, $\gamma > 0$ such that

(i) a distribution $S$ given by (9.1) is regular on $\Omega$ with kernel $K$ satisfying

$$(9.3) \qquad |K(x, y)| \leq C/\rho(x - y),$$

(ii) If $(x, y) \in \Omega$ and $\rho(x' - x) \leq \rho(x - y)/b^{2\omega}$ then

$$(9.4) \qquad |K(x', y) - K(x, y)| \leq C \frac{\rho(x' - x)^\gamma}{\rho(x - y)^{1+\gamma}},$$

(iii) If $(x, y) \in \Omega$ and $\rho(y' - y) \leq \rho(x - y)/b^{2\omega}$ then

$$(9.5) \qquad |K(x, y') - K(x, y)| \leq C \frac{\rho(y' - y)^\gamma}{\rho(x - y)^{1+\gamma}},$$

(iv) $T$ extends to a continuous linear operator on $L^2(\mathbb{R}^n)$ with $||T|| \leq C$.

A few remarks are needed. It can be shown that operators in the above class are bounded from $L^1$ to weak-$L^1$. This follows from the general theory of Calderón-Zygmund operators defined on arbitrary spaces of homogeneous type, see [St2, Chapter 1]. By the Marcinkiewicz Interpolation Theorem they are bounded from $L^p$ into $L^p$ for $1 < p \leq 2$. By taking duals they are also bounded for $2 \leq p < \infty$, hence in the range $1 < p < \infty$. To obtain these conclusions, conditions (ii) and (iii) in Definition 9.1 can be relaxed to the weaker Hörmander integral conditions [Hö].

The condition (iv) immediately implies that $T : \mathcal{S} \to \mathcal{S}'$ is bounded. We adopt this seemingly redundant definition because in many situations the boundedness of $T$ on $L^2$ is not automatic. The operators satisfying all the conditions in Definition 9.1 except (iv) are sometimes called *generalized Calderón-Zygmund operators*. The famous $T(1)$ Theorem of David and Journé [DJ] gives a necessary and sufficient condition for such $T$ to be bounded on $L^2$. This result was further generalized by David, Journé and Semmes [DJS] to the setting of spaces of homogeneous type and to the more general $T(b)$ Theorem, where $b$ belongs to the special class of admissible functions, see also [Dv].

Since we are interested in boundedness results on $H^p$ spaces, 0 require much stricter conditions on the kernel $K(x, y)$ than those (iii). These conditions are known in the case when the dilation $A$ over $\mathbb{R}$ and they involve the appropriate decay of the directional with respect to the eigenvectors of $A$, see [St2, Chapter 13.5] and 6]. For general dilations we must be more careful, requiring smoo which hold uniformly after rescaling to the scale zero.

DEFINITION 9.2. We say that $T$ is a *Calderón-Zygmund opera* $T$ satisfies Definition 9.1 with $K(x, y)$ in the class $C^m$ as a functio require that there exists a constant $C$ such that for every $(x, y) \in \mathcal{S}$

$$(9.6) \qquad |\partial_y^\alpha [K(\cdot, A^k \cdot)](x, A^{-k}y)| \leq C/\rho(x-y) = Cb^{-k} \qquad \text{for } |\alpha|$$

where $k \in \mathbb{Z}$ is the unique integer such that $x - y \in B_{k+1} \setminus B_k$. I $\partial_y^\alpha[K(\cdot, A^k \cdot)](x, A^{-k}y)$ means $\partial_y^\alpha \tilde{K}(x, A^{-k}y)$, where $\tilde{K}(x, y) = K(x,$

In short, we say that $T$ is (CZ-$m$) and the smallest $C$ fulfilli (i)–(iv) of Definition 9.1 and (9.6) is denoted by $||T||_{(m)}$.

EXAMPLE. In the case when $A$ is an isotropic dilation, in particu multiple of the identity, then (9.6) takes the familiar form

$$|\partial_y^\alpha K(x, y)| \leq C|x - y|^{-n-|\alpha|} \qquad \text{for } |\alpha| \leq m.$$

More generally, suppose $A$ is a diagonal matrix with diagonal terms where $a_1, \ldots, a_n > 0$ and $a = a_1 + \ldots + a_n$. Then $\rho$ given by $\rho(x_1$ $\max_{1 \leq i \leq n} |x_i|^{a/a_i}$ is a quasi-norm associated with $A$. Pick any $x, y$ $|\det A|^k \leq \rho(x-y) < |\det A|^{k+1}$ for some $k \in \mathbb{Z}$. Since

$$|\partial_y^\alpha[K(\cdot, A^k \cdot)](x, A^{-k}y) = e^{k(\alpha_1 a_1 + \ldots + \alpha_n a_n)}|\partial_y^\alpha K(x, y)| \leq C\rho(x-y)$$

where $\alpha = (\alpha_1, \ldots, \alpha_n)$, the condition (9.6) takes a more familiar form

$$|\partial_y^\alpha K(x, y)| \leq C\rho(x-y)^{-1-(\alpha_1 a_1 + \ldots + \alpha_n a_n)/a} \qquad \text{for } |\alpha| \leq m,$$

see [St2, Chapter 13.5]. However, if $A$ has some non-trivial blocks in it decomposition then (9.6) does not have a more explicit form due to the co of the action of $A$ on $\mathbb{R}^n$.

The following basic fact provides a sufficient condition for a function t to $H^p$.

LEMMA 9.3. *Let $(p, q, s)$ be admissible and $\delta > \max(1/p, s \ln \lambda_+ / \ln$ Suppose that $f$ is a measurable function on $\mathbb{R}^n$ such that for some constant we have*

$$(9.7) \qquad \left( \frac{1}{|B_k|} \int_{B_k} |f(x)|^q dx \right)^{1/q} \leq C|B_k|^{-1/p} \qquad \text{for some } k \in \mathbb{Z},$$

$$(9.8) \qquad |f(x)| \leq C|B_k|^{-1/p} \rho(A^{-k}x)^{-\delta} \qquad \text{for } x \in B_k^{\mathsf{c}},$$

$$(9.9) \qquad \int_{\mathbb{R}^n} f(x) x^\alpha dx = 0 \qquad \text{for } |\alpha| \leq s.$$

Since $\|\mathbf{1}_{B_j} D_{A^{-j}} Q_\alpha\|_\infty = \|Q_\alpha \mathbf{1}_{B_0}\|_\infty \le C_1$ for all $|\alpha| \le s$, by (9.9) and (9.11) we have

$$\|\mathbf{1}_{B_j} \pi_{B_j} f\|_\infty \le C_1 b^{-j} \sum_{|\alpha| \le s} \left| \int_{B_j^c} f(x) \overline{Q_\alpha(A^{-j}x)} dx \right|.$$

Since $|Q_\alpha(x)| \le C_2 |x|^s$ for $x \in B_0^c$ for some constant $C_2 > 0$

$$\left| \int_{B_j^c} f(x) \overline{Q_\alpha(A^{-j}x)} dx \right| \le C_2 \int_{B_j^c} |f(x)| |A^{-j}x|^s dx$$

$$\le C_2 \int_{B_j^c} C|B_k|^{-1/p} \rho(A^{-k}x)^{-\delta} \rho(A^{-j}x)^{s \ln \lambda_+ / \ln b} dx$$

$$= C_2 C b^{-k/p} b^{\delta(k-j)} \int_{B_j^c} \rho(A^{-j}x)^{-\delta + s \ln \lambda_+ / \ln b} dx$$

$$\le C_2 C b^{-k/p} b^{\delta(k-j)} b^j \int_{B_0^c} \rho(x)^{-\delta + s \ln \lambda_+ / \ln b} dx = C_3 b^{j(1-1/p)} b^{(\delta - 1/p)(k-j)}.$$

Inserting the last three inequalities into (9.13) we conclude that

$$\|g_{j+1} - g_j\|_\infty \le C_4 b^{-(j+1)/p} b^{(\delta - 1/p)(k-j)} \qquad \text{for } j \ge k,$$

for some constant $C_4$. Since $g_j$'s have vanishing moments up to order $s$, $g_{j+1} - g_j$ is a $\kappa_j$ multiple of a $(p, \infty, s)$-atom $a_j$ supported on $B_{j+1}$, where $g_{j+1} - g_j = \kappa_j a_j$, $\kappa_j = C_4 b^{(\delta - 1/p)(k-j)}$. By (9.12)

$$\|f\|_{H^p_{q,s}} \le \left( (C_0 C)^p + \sum_{j=k}^\infty |\kappa_j|^p \right)^{1/p} = \left( (C_0 C)^p + C_4^p \sum_{j=k}^\infty b^{(p\delta - 1)(k-j)} \right)^{1/p} = C',$$

which ends the proof of the lemma. $\square$

REMARK. Since translations are isometries in $H^p$ we can immediately generalize Lemma 9.3 to functions centered at arbitrary $x_0 \in \mathbb{R}^n$. Conditions (9.7) and (9.8) can be substituted then by

(9.14) $\qquad \left( \dfrac{1}{|B_k|} \displaystyle\int_{x_0 + B_k} |f(x)|^q dx \right)^{1/q} \le C |B_k|^{-1/p},$

(9.15) $\qquad |f(x)| \le C |B_k|^{-1/p} \rho(A^{-k}(x - x_0))^{-\delta} \qquad$ for $x \in x_0 + B_k^c$.

A function $f$ satisfying (9.9), (9.14), and (9.15) with $C = 1$ is refferred to as a *molecule* localized around the dilated ball $x_0 + B_k \in \mathcal{B}$. Hence, a molecule satisfies all the properites of an atom as in Definition 4.1 with the exception of compact support condition, which is replaced by a suitable decay condition (9.15). Therefore, Lemma 9.3 says that, for fixed $(p, q, s)$ and decay exponent $\delta$, every molecule $f$ belongs to $H^p$ with the $H^p$ norm bounded by some constant depending only on $(p, q, s)$ and $\delta$. Moreover, by examining the arguments of Lemma 9.3, one can easily see that, for fixed $0 < p \le 1$, this constant depends only on $\delta$.

We also remark that our definition of molecule is more restrictive than what normally is understood as a molecule. For properties of molecules we refer the interested reader to [BS, CW2, GR, TW].

Our next goal is to show that Calderón-Zygmund operators map atoms into molecules. Generally, we can not expect this unless we also assume that our operator preserves vanishing moments. The precise meaning of this is given in Definition 9.4.

DEFINITION 9.4. We say that a Calderón-Zygmund operator $T$ of order $m$ satisfies $T^*(x^\alpha) = 0$ for all $|\alpha| \leq s$, where $s < m \ln \lambda_- / \ln \lambda_+$, if for every $f \in L^2$ with compact support and $\int x^\alpha f(x) dx = 0$ for $|\alpha| < m$, we also have $\int x^\alpha T f(x) dx = 0$ for all $|\alpha| \leq s$.

Note that we require that $s < m \ln \lambda_- / \ln \lambda_+$ in Definition 9.4 to guarantee that the integrals $\int x^\alpha T f(x) dx$ are well defined for all $|\alpha| \leq s$. Indeed, it follows from the proof of Lemma 9.5 that $|Tf(x)| = O(\rho(x)^{-1-m\ln\lambda_-/\ln b})$ as $|x| \to \infty$, and hence $|Tf(x)| = O(\rho(x)^{-1}|x|^{-m\ln\lambda_-/\ln\lambda_+})$ as $|x| \to \infty$. We also remark that Definition 9.4 overlaps with analogous property in the isotropic setting investigated by Meyer in [MC, Chapter 7.4]. Furthermore, the condition $T^*(x^\alpha) = 0$ is automatically satisfied when $T$ is a convolution singular integral operator, which explains why this condition does not appear in this simpler situation, see [GR, Chapter III.7].

LEMMA 9.5. *Suppose that $(p, 2, s)$ is admissible and an integer $m$ satisfies $m > \max((1/p - 1) \ln b / \ln \lambda_-, s \ln \lambda_+ / \ln \lambda_-)$. Assume $T$ is (CZ-m) and $T^*(x^\alpha) = 0$ for $|\alpha| \leq s$. Then there exists a constant $C$, depending only on the Calderón-Zygmund norm $\|T\|_{(m)}$ of $T$, such that $\|Ta\|_{H^p_{2,s}} \leq C$ for every $(p, 2, m - 1)$-atom $a$.*

PROOF. Suppose a $(p, 2, m - 1)$-atom $a$ is supported in the ball $x_0 + B_k$. We estimate $Ta$ separately around and away from the support of an atom $a$.

$$(9.16) \qquad \int_{x_0+B_{k+\omega}} |Ta(x)|^2 dx \leq \int_{\mathbb{R}^n} |Ta(x)|^2 dx \leq C\|a\|_2^2 \leq |B|^{1-2/p}.$$

Suppose $x \in x_0 + B_{k+l+\omega+1} \setminus B_{k+l+\omega}$ for some $l \geq 0$ and $y \in x_0 + B_k$. Then $x - y \in B_{k+l+2\omega+1}$ but $x - y \notin B_{k+l}$. Hence, by (9.6) and the chain rule

$$(9.17) \qquad |\partial_y^\alpha [K(\cdot, A^{k+l}\cdot)](x, A^{-k-l}y)| \leq C'/\rho(x-y) = C'b^{-k-l} \qquad \text{for } |\alpha| \leq m,$$

where $C'$ depends only on the constant $C$ in (9.6). Away from the support of the atom $a$, we estimate $Ta$ by

$$(9.18) \qquad |Ta(x)| = \left|\int_{x_0+B_k} K(x,y)a(y)dy\right| = \left|\int_{x_0+B_k} \tilde{K}(x, A^{-k-l}y)a(y)dy\right|,$$

where $\tilde{K}(x, y) = K(x, A^{k+l}y)$. Now we expand $\tilde{K}(x, y)$ into the Taylor polynomial of degree $m - 1$ (only in $y$ variable) at the point $(x, A^{-k-l}x_0)$. That is,

$$(9.19) \qquad \tilde{K}(x, y') = \sum_{|\alpha| \leq m-1} \frac{\partial_y^\alpha \tilde{K}(x, A^{-k-l}x_0)}{\alpha!}(y' - A^{-k-l}x_0)^\alpha + R_m(y').$$

Here we think that $y' = A^{-k-l}y$ and that $y$ ranges over $x_0 + B_k$ as in (9.18). Since we are going to apply (9.19) for $y' \in A^{-k-l}x_0 + B_{-l}$, the remainder $R_m$ satisfies

$$(9.20) \qquad \begin{aligned} |R_m(y')| &\leq \tilde{C} \sup_{z \in A^{-k-l}x_0 + B_{-l}} \sup_{|\alpha|=m} |\partial_y^\alpha \tilde{K}(x,z)||y' - A^{-k-l}x_0|^m \\ &\leq C'b^{-k-l} \sup_{z \in B_{-l}} |z|^m, \end{aligned}$$

because the partial derivatives of $\tilde{K}(x,y)$ in the variable $y$ of order $m$ satisfy (9.17). Combining (9.18), (9.19), and (9.20) and using the moment condition of atoms we have

$$|Ta(x)| = \left| \int_{x_0+B_k} R_m(A^{-k-l}y)a(y)dy \right| \leq \int_{x_0+B_k} |R_m(A^{-k-l}y)a(y)|dy$$

$$\leq C'b^{-k-l} \sup_{z \in B_{-l}} |z|^m \int_{x_0+B_k} |a(y)|dy \leq C'b^{-k-l}(c\lambda_-^{-l})^m b^{k-k/p}$$

$$= C''b^{-k/p}b^{-l}\lambda_-^{-lm} = C''b^{-k/p}b^{-l\delta},$$

where $\delta = m \ln \lambda_- / \ln b + 1$. Therefore, $Ta$ satisfies (9.14) and (9.15). Furthermore, we have $T^*(x^\alpha) = 0$ for $|\alpha| \leq s$ meaning $Tf$ has vanishing moments up to order $s$ whenever $f \in L^2$ with compact support has vanishing moments up order $m-1$, i.e., $f$ is a multiple of a $(p, 2, m-1)$-atom. By Lemma 9.3 this implies that there is a constant $C$ independent of $a$ so that $||Ta||_{H^p_{2,s}} \leq C$. □

Lemma 9.5 strongly suggests that any Calderón-Zygmund operator $T$ of order $m$ extends to a bounded operator on the Hardy space $H^p$, where $m$ is sufficiently large and depends on $0 < p \leq 1$. This requires a careful proof since potentially there could be a problem with the well-definedness of $T$ on $H^p$, due to the non-uniqueness of atomic decompositions. This problem is similar to the one we encountered when dealing with the duals of $H^p$. To overcome this difficulty we need to use an approximation of a given Calderón-Zygmund operator $T$ of order $m$ by a sequence of (CZ-$m$) operators with nonsingular kernels.

In the following theorem we impose the same constraints on integers $m$ and $s$ as in Lemma 9.5.

THEOREM 9.6. *Suppose $T$ is a (CZ-$m$) with kernel $K(x,y)$. Then there is a sequence $(T_i)_{i \in \mathbb{Z}}$ of uniformly bounded (CZ-$m$), i.e., for any $i \in \mathbb{Z}$, $||T_i||_{(m)} \leq \tilde{C}$, with nonsingular $C^\infty$ kernels $K_i(x,y)$ such that*

(9.21) $$T_i f(x) = \int K_i(x,y)f(y)dy \quad \text{for } x \in \mathbb{R}^n,$$

*where $f \in L^2$ has compact support, and*

(9.22) $$T_i f \to Tf \quad \text{in } L^2 \text{ as } i \to -\infty \quad \text{for all } f \in L^2.$$

*Furthermore, if $T^*(x^\alpha) = 0$ for $|\alpha| \leq s$ then also $(T_i)^*(x^\alpha) = 0$ for all $|\alpha| \leq s$ and $i \in \mathbb{N}$.*

One possible approximation technique for Calderón-Zygmund operators involves the truncation of the kernel. Given a $C^\infty$ function $\varphi$ such that $\varphi(x) = 0$ for $x \in B_{-1}$, $\varphi(x) = 1$ for $x \in (B_0)^c$ we can define kernels $K_i$ by $K_i(x,y) = K(x,y)\varphi(A^{-i}(x-y))$. By adapting arguments of Meyer in [MC] it is possible to show, though it is not automatic, that the family of the corresponding operators $(T_i)$ is (CZ-$m$) with uniform constants. Furthermore, $T_{i_j}$'s (after taking a subsequence) converge weakly to $M_h + T$ as $i_j \to -\infty$, where $M_h$ denotes the operator of multiplication by a function $h \in L^\infty$, see [MC, Chapter 7, Proposition 3]. However, if $K(x,y)$ is not a convolution kernel then in general the $T_i$'s might not preserve vanishing moments. Therefore, this approximation is not suitable for showing boundedness of $T: H^p \to H^p$. It can be used, though, to show boundedness of $T: H^p \to L^p$.

Instead we are going to use an approximation based on smoothing by a convolution with a compactly supported smooth function $\varphi$. Indeed, suppose $\varphi$ is $C^\infty$, $\operatorname{supp} \varphi \subset B_0$ and $\int \varphi = 1$. For any $i \in \mathbb{Z}$ define a convolution operator $R_i: \mathcal{S}' \to \mathcal{S}'$, by

$$(9.23) \qquad R_i f = f * \varphi_i.$$

By Lemma 6.6 for every $0 < p < \infty$ we have

$$(9.24) \qquad \|R_i f\|_{H^p} \leq C \|f\|_{H^p},$$

and by Theorem 6.8,

$$(9.25) \qquad \lim_{i \to -\infty} \|R_i f - f\|_{H^p} = 0 \quad \text{for every } f \in H^p.$$

We are now ready to present the proof of Theorem 9.6.

PROOF OF THEOREM 9.6. We define operators $T_i$ by $T_i = R_i T R_i$, where $R_i$ is given by (9.23). Since $T_i: \mathcal{S} \to \mathcal{S}'$ it has a kernel $K_i \in \mathcal{S}'(\mathbb{R}^n \times \mathbb{R}^n)$ by the Schwartz Kernel Theorem. We claim that $K_i$ is a regular distribution which is identified with the function $K_i$ given by

$$(9.26) \qquad K_i(x, y) = \langle T(\tau_y \varphi_i), \tau_x \tilde{\varphi}_i \rangle,$$

where $\tilde{\varphi}_i(z) = \varphi_i(-z)$. Indeed, recall that for any $f \in \mathcal{S}'$, $R_i f = f * \varphi_i$ is a regular distribution identified with $R_i f(x) = f(\tau_x \tilde{\varphi}_i)$. For any $\psi \in \mathcal{S}$, $R_i T R_i(\psi)$ is also regular distribution and

$$(R_i \circ T \circ R_i)\psi(x) = \langle T R_i(\psi), \tau_x \tilde{\varphi}_i \rangle = \langle T(\varphi_i * \psi), \tau_x \tilde{\varphi}_i \rangle$$
$$= \left\langle T\left(\int \psi(y) \varphi_i(\cdot - y) dy\right), \tau_x \tilde{\varphi}_i \right\rangle = \left\langle \int \psi(y) T(\tau_y \varphi_i)(\cdot) dy, \tau_x \tilde{\varphi}_i \right\rangle$$
$$= \int \psi(y) \langle T(\tau_y \varphi_i), \tau_x \tilde{\varphi}_i \rangle dy.$$

The next to last equality is justified by approximating the integral in $L^2$ by finite linear combinations of functions $\varphi(\cdot - y)$ for $y \in \mathbb{R}^n$. This shows (9.26). Moreover, if $x - y \notin 2B_i \subset B_{i+\omega}$ then supports of $\tau_y \varphi_i$ and $\tau_x \tilde{\varphi}_i$ are disjoint and

$$(9.27) \qquad \begin{aligned} K_i(x, y) &= \langle T(\tau_y \varphi_i), \tau_x \tilde{\varphi}_i \rangle = \int_{\mathbb{R}^n} \int_{\mathbb{R}^n} K(u, v) \varphi_i(v - y) dv \varphi_i(x - u) du \\ &= \int_{\mathbb{R}^n} \int_{\mathbb{R}^n} K(x - u, y + v) \varphi_i(u) \varphi_i(v) du dv. \end{aligned}$$

Fix $(x_0, y_0) \in \Omega$ and suppose that $x_0 - y_0 \in B_{l+2\omega+1} \setminus B_{l+2\omega}$ for some $l \geq i$. Since $u$ and $v$ range over $B_i$ in (9.27) $x_0 - u - (y_0 + v) = x_0 - y_0 - u - v \in B^c_{l+2\omega} + B_{i+\omega} \subset B^c_{l+\omega}$. Also $x_0 - y_0 - u - v \in B_{l+3\omega+1}$. Note that for $|\alpha| \leq m$

$$(9.28) \quad \partial_y^\alpha[K_i(\cdot, A^{l+2\omega} \cdot)](x, y) = \int\int \partial_y^\alpha[K(\cdot - u, A^{l+2\omega} \cdot + v)](x, y) \varphi_i(u) \varphi_i(v) du dv,$$

by moving the differentiation inside the integral. By the chain rule and (9.6) there is a constant $C'$ so that

$$(9.29) \qquad |\partial_y^\alpha[K(\cdot, A^{l+2\omega} \cdot)](x_0 - u, A^{-l-2\omega}(y_0 + v))| \leq C' b^{-l-2\omega}$$

for all $|\alpha| \leq m$ and $u, v \in B_i$. Combining (9.28) and (9.29) we obtain

$$\text{(9.30)} \quad |\partial_y^\alpha[K_i(\cdot, A^{l+2\omega}\cdot)](x_0, A^{-l-2\omega}y_0)| \leq C'b^{-l-2\omega} = C'\rho(x_0 - y_0)^{-1}.$$

Suppose next that $x_0 - y_0 \in B_{l+1} \setminus B_l$ for some $l < i + 2\omega$. We claim that there is a constant $C'$ such that

$$\text{(9.31)} \quad |\partial_y^\alpha[K_i(\cdot, A^l\cdot)](x, y)| \leq C'b^{-i} \leq C'b^{2\omega}\rho(x_0 - y_0)^{-1}$$

for all $(x, y) \in \mathbb{R}^n \times \mathbb{R}^n$ and $|\alpha| \leq m$. In particular, by choosing $(x, y) = (x_0, A^{-l}y_0)$ we obtain the estimate (9.6) for the kernel $K_i$. To see (9.31) for $\alpha = 0$ we use (9.26) and the Cauchy-Schwarz inequality

$$|K_i(x, y)| \leq ||T|| ||\varphi_i||_2^2 = ||T|| ||\varphi||_2^2 b^{-i} = C'b^{-i}.$$

For $\alpha \neq 0$ we need an additional argument. Define the mapping $H: \mathbb{R}^n \to L^2(\mathbb{R}^n)$ by $H(y) = \tau_{A^ly}\varphi_i$. Clearly, $H$ is a $C^\infty$ function on $\mathbb{R}^n$ with values in the Hilbert space $L^2(\mathbb{R}^n)$. Moreover, $T \circ H$ is also $C^\infty$ and $\partial^\alpha(T \circ H) = T \circ \partial^\alpha H$ for any multi-index $\alpha$, since bounded linear maps commute with the differentiation. Consider also the function $h(x, y) = \tau_{A^ly}\varphi_i(x)$. For fixed $y \in \mathbb{R}^n$, $\operatorname{supp} h(\cdot, y) \subset A^ly + B_i$. Also for $|\alpha| \leq m$

$$|\partial_y^\alpha h(x, y)| = b^{-i}|\partial_y^\alpha \varphi(A^{-i}(x - A^l\cdot))(y)| = |b^{-i}\partial_y^\alpha \varphi(A^{-i}x - A^{l-i}\cdot)(y)| \leq C''b^{-i},$$

by the chain rule since $\varphi$ is $C^\infty$ with compact support and $l - i < 2\omega$. Therefore, for any $y \in \mathbb{R}^n$,

$$\text{(9.32)} \quad ||\partial^\alpha H(y)||_2^2 = \int_{\mathbb{R}^n} |\partial_y^\alpha h(x, y)|^2 dx = \int_{A^ly + B_i} |\partial_y^\alpha h(x, y)|^2 dx \leq C''b^{-i}.$$

Therefore, by (9.32) and the Cauchy-Schwarz inequality

$$|\partial_y^\alpha[K_i(\cdot, A^l\cdot)](x, y)| = |\langle \partial^\alpha[T \circ H](y), \tau_x\tilde{\varphi}_i\rangle| \leq ||T|| |\langle \partial^\alpha H(y), \tau_x\tilde{\varphi}_i\rangle|$$
$$\leq ||T|| ||\partial_y^\alpha H(y)||_2 ||\varphi_i||_2 \leq C''||T|| ||\varphi||_2 b^{-i},$$

which shows (9.31). The estimates (9.30) and (9.31) cover the whole range of $(x_0, y_0) \in \Omega$ and they imply that the $T_i$'s are (CZ-$m$) with norms $||T||_{(m)}$ independent of $i \in \mathbb{Z}$. Clearly, $T_i$'s are uniformly bounded on $L^2$ since the $R_i$'s are uniformly bounded on $L^2$ by (9.24). Moreover, (9.22) holds by (9.25). Moreover, $K_i(x, y)$ is a smooth kernel function on $\mathbb{R}^n \times \mathbb{R}^n$ by (9.26). Hence, $T_i$ has a nonsingular kernel $K_i(x, y)$ satisfying (9.21) initially for $f \in \mathcal{S}$. By a density argument, (9.21) holds for all $f \in L^2$ with compact support.

Finally, we ought to show that given a function $f \in L^2$ with compact support and $\int f(x)x^\alpha = 0$ for $|\alpha| \leq m - 1$ we have $\int T_if(x)x^\alpha = 0$ for $|\alpha| \leq s$. This is a consequence of Lemma 9.7. Indeed, $R_if$ is also in $L^2$ with vanishing moments up to order $m-1$, hence it is a multiplicity of some $(p, 2, m-1)$-atom. By Lemma 9.5, $TR_if$ satisfies the decay estimate (9.15) and has vanishing moments up to order $s$. Therefore, by Lemma 9.7, $R_iTR_if$ has also vanishing moments up to order $s$. This ends the proof of Theorem 9.6. $\square$

LEMMA 9.7. *Suppose $f$ satisfies $f(x)(1 + |x|^{\tilde{m}}) \in L^1$ and $\int f(x)x^\alpha dx = 0$ for $|\alpha| \leq \tilde{m}$. Then we also have $\int R_if(x)x^\alpha dx = 0$ for $|\alpha| \leq \tilde{m}$.*

PROOF. For any $|\alpha| \le \tilde{m}$ by the Fubini Theorem

$$\int_{\mathbb{R}^n} x^\alpha R_i f(x) dx = \int_{\mathbb{R}^n} \int_{\mathbb{R}^n} x^\alpha f(x-y) \varphi_i(y) dy dx$$
$$= \int_{\mathbb{R}^n} \int_{\mathbb{R}^n} (x+y)^\alpha f(x) \varphi_i(y) dx dy = 0,$$

because the integrand belongs to $L^1(\mathbb{R}^n \times \mathbb{R}^n)$ since the support of $\varphi$ is bounded. □

We are now ready to prove the main result of this section.

THEOREM 9.8. *Suppose $T$ is a Calderón-Zygmund operator of order $m$. If $p$ satisfies*

$$(9.33) \qquad 0 \le 1/p - 1 < \frac{(\ln \lambda_-)^2}{\ln b \ln \lambda_+} m,$$

*then $T$ extends to the continuous linear operator $T: H^p(\mathbb{R}^n) \to H^p(\mathbb{R}^n)$ provided $T^*(x^\alpha) = 0$ for $|\alpha| \le s = \lfloor (1/p - 1) \ln b / \ln \lambda_- \rfloor$.*

PROOF. Note that (9.33) guarantees that $p$ and $m$ satisfy the assumptions of Lemma 9.5 and Theorem 9.6, i.e., $m > \max((1/p-1)\ln b/\ln \lambda_-, s \ln \lambda_+ / \ln \lambda_-)$. Let $(T_i)_{i\in\mathbb{Z}}$ be the sequence of operators with kernels $K_i(x,y)$ given by Theorem 9.6. We claim that for any $(p, 2, m-1)$-atom $a$ we have

$$(9.34) \qquad T_i a \to Ta \quad \text{in } H^p \text{ as } i \to -\infty.$$

Indeed, if $a$ is supported in the ball $x_0 + B_k$ then $R_i a$ is supported in $x_0 + B_{k+\omega}$ for $i \le k$. Since $\|a - R_i a\|_2 \to 0$ as $i \to -\infty$, therefore $a - R_i a$ is a $\kappa_i$ multiple of some $(p, 2, m-1)$ atom for $i \le k$ by Lemma 9.7. Furthermore, $\kappa_i \to 0$ as $i \to -\infty$. By Lemma 9.5, $\|T(a - R_i a)\|_{H^p_{2,s}} \to 0$ as $i \to \infty$. By (9.24) and (9.25),

$$\|Ta - R_i T R_i a\|^p_{H^p} \le \|Ta - R_i Ta\|^p_{H^p} + \|R_i(Ta - TR_i a)\|^p_{H^p} \to 0 \quad \text{as } i \to -\infty,$$

which shows (9.34).

The kernel function $K_i(x,y)$ of $T_i$, given by (9.26), is $C^\infty$. Furthermore, by the Cauchy-Schwarz inequality all partial derivatives of $K_i(x,y)$ of order $\le N$ are uniformly bounded for any natural number $N$. By taking $N \ge (1/p-1)\ln b/\ln \lambda_-$ we conclude by Proposition 8.7 that $K_i(x, \cdot)$ belongs to the Campanato space $C^l_{\infty, s}$ with $l = 1/p - 1$ for any fixed $x \in \mathbb{R}^n$. Furthermore,

$$(9.35) \qquad \|K_i(x, \cdot)\|_{C^{1/p-1}_{\infty,s}} \le C(i) \qquad \text{for all } x \in \mathbb{R}^n,$$

and for some constant $C = C(i)$ depending only on $i \in \mathbb{Z}$.

Take any $f \in L^2$ with compact support and vanishing moments up to order $m-1$, that is $f \in \Theta^2_{m-1}$. Let $f = \sum_{j\in\mathbb{N}} \kappa_j a_j$ be an atomic decomposition of $f$ into $(p, 2, m-1)$ atoms with $\sum_{j\in\mathbb{N}} |\kappa_j|^p \le 2\|f\|^p_{H^p_{2,m-1}}$. By the duality Theorem 8.3, $K_i(x, \cdot)$ defines a bounded functional on $H^p$ and by (8.18) and (9.35),

$$(9.36) \qquad T_i f(x) = \sum_{j\in\mathbb{N}} \kappa_j T_i a_j(x) \qquad \text{for every } x \in \mathbb{R}^n.$$

Furthermore, the convergence in (9.36) is uniform on $\mathbb{R}^n$ by (8.15) and (9.35), and hence the series in (9.36) converges in $\mathcal{S}'$. By Theorem 9.6, $T_i$'s are uniformly bounded (CZ-$m$) and hence by Lemma 9.5, $||T_i a_j||_{H^p} \leq C$ for some constant $C$ independent of $i$ and $j$. Therefore,

$$(9.37) \qquad T_i f = \sum_{j \in \mathbb{N}} \kappa_j T_i a_j \qquad \text{convergence in } H^p,$$

Moreover,

$$||T_i f||_{H^p}^p \leq \sum_{j \in \mathbb{N}} |\kappa_j|^p ||T_i a_j||_{H^p}^p \leq 2 C^p ||f||_{H^p_{2,m-1}}^p.$$

Combining this with (9.34), (9.37), and letting $i \to -\infty$, we obtain

$$(9.38) \qquad ||Tf||_{H^p} \leq C' ||f||_{H^p} \qquad \text{for } f \in \Theta^2_{m-1}$$

for some constant $C'$ independent of $f$. Moreover,

$$(9.39) \qquad Tf = \sum_{j \in \mathbb{N}} \kappa_j T a_j \qquad \text{convergence in } H^p.$$

Since $\Theta^2_{m-1}$ is a dense subspace of $H^p$, $T$ extends uniquely to a bounded operator in $H^p$ by (9.38). $\square$

As a consequence we conclude that for any $f \in H^p$ with $f = \sum_{j \in \mathbb{N}} \kappa_j a_j$, where $\sum_{j \in \mathbb{N}} |\kappa_j|^p < \infty$ and $a_j$'s are $(p, 2, s)$-atoms, we can compute $Tf$ by applying the formula (9.39).

In the case when $T$ does not necessarily satisfy $T^*(x^\alpha) = 0$ for $|\alpha| \leq s$, we still have a boundedness result which is analogous to Theorem 9.8.

THEOREM 9.9. *Suppose $T$ is a Calderón-Zygmund operator of order $m$. If $p$ satisfies (9.33) then $T$ extends to the continuous linear operator $T : H^p(\mathbb{R}^n) \to L^p(\mathbb{R}^n)$.*

PROOF. The proof follows along the lines of the proof of Theorem 9.8 with the exception that (9.34) may not hold, since $Ta$ may not even belong to $H^p$ for a $(p, 2, m-1)$ atom $a$. Nevertheless, $Ta(x)$ satisfies the same size estimates (9.14) and (9.15) as in Lemma 9.5. Therefore, $||Ta||_{L^p} \leq C$ for some constant $C$ depending on $||T||_{(m)}$, but independent of an atom $a$. Moreover, (9.36) still holds for $f \in \Theta^2_{m-1}$ and consequently $||T_i f||_{L^p} \leq C ||f||_{H^p}$ for all $i \in \mathbb{Z}$ and $f \in \Theta^2_{m-1}$. By (9.22) we have $T_i f(x) \to Tf(x)$ for a.e. $x \in \mathbb{R}^n$ as $i \to -\infty$ (possibly after taking a subsequence). Therefore, by Fatou's Lemma $||Tf||_{L^p} \leq C ||f||_{H^p}$ for all $f \in \Theta^2_{m-1}$. Therefore, $T$ has a unique extension from $H^p$ into $L^p$, which concludes the proof of Theorem 9.9. $\square$

## 10. Classification of dilations

In this section we investigate the question when two different dilations generate the same anisotropic Hardy space $H^p$ for some $0 < p \leq 1$. We give a necessary and sufficient condition for this to happen in terms of the spectral properties of these dilations. We start with two basic lemmas.

LEMMA 10.1. *Suppose $D$ is an $n \times n$ matrix. If $||D|| \leq c_1$ and $|\det D| \geq c_2$ for some $c_1, c_2 > 0$ then there is a constant $c_3 = c_3(c_1, c_2) > 0$ independent of $D$ such that $|Dz| \geq c_3|z|$ for all $z \in \mathbb{R}^n$. Similarly, if $|Dz| \geq c_1|z|$ for all $z \in \mathbb{R}^n$ and $|\det D| \leq c_2$ for some $c_1, c_2 > 0$ then there is a constant $c_3 = c_3(c_1, c_2) > 0$ independent of $D$ such that $||D|| \leq c_3$.*

LEMMA 10.2. *Suppose we have two dilations $A_1$ and $A_2$ on $\mathbb{R}^n$. Let $\rho_1$ and $\rho_2$ be the quasi-norms associated to $A_1$ and $A_2$, respectively. Then $\rho_1$ and $\rho_2$ are equivalent if and only if*

$$(10.1) \qquad \inf_{z \in \mathbb{R}^n \setminus \{0\}} \inf_{k \in \mathbb{Z}} |A_1^k z|/|A_2^{\lfloor \epsilon k \rfloor} z| > 0,$$

*or*

$$(10.2) \qquad \sup_{z \in \mathbb{R}^n \setminus \{0\}} \sup_{k \in \mathbb{Z}} |A_1^k z|/|A_2^{\lfloor \epsilon k \rfloor} z| < \infty,$$

*where*

$$(10.3) \qquad \varepsilon = \varepsilon(A_1, A_2) = \ln|\det A_1|/\ln|\det A_2|.$$

PROOF. Note that for every $k \in \mathbb{Z}$

$$(10.4) \quad \begin{aligned} 1 &= |\det A_1|^k |\det A_2|^{-\epsilon k} \leq |\det A_1|^k |\det A_2|^{-\lfloor \epsilon k \rfloor} \\ &= |\det(A_1^k A_2^{-\lfloor \epsilon k \rfloor})| \leq |\det A_1|^k |\det A_2|^{-\epsilon k + 1} = |\det A_2|. \end{aligned}$$

By Lemma 10.1 applied to $A_1^k A_2^{-\lfloor \epsilon k \rfloor}$ we obtain (10.1) is equivalent to (10.2).

Assume that (10.1) and hence (10.2) holds. Let $c$ and $d$ denote the values of (10.1) and (10.2), respectively. Fix $r > 0$ so that for every $z \in \mathbb{R}^n \setminus \{0\}$, and $i = 1, 2$, there exists $k \in \mathbb{Z}$ such that $1 \leq |A_i^k z| \leq r$. Clearly, $r = \max(||A_1||, ||A_2||)$ works. Denote

$$c_1 = \inf\{\rho_1(z) : 1 \leq |z| \leq r\}, \qquad d_1 = \sup\{\rho_1(z) : 1 \leq |z| \leq r\},$$
$$c_2 = \inf\{\rho_2(z) : 1/d \leq |z| \leq r/c\}, \qquad d_2 = \sup\{\rho_2(z) : 1/d \leq |z| \leq r/c\}.$$

Fix $x \in \mathbb{R}^n \setminus \{0\}$ and choose $k \in \mathbb{Z}$ such that $1 \leq |A_1^k x| \leq r$. Clearly

$$(10.5) \qquad |\det A_1|^{-k} c_1 \leq \rho_1(x) \leq |\det A_1|^{-k} d_1.$$

By (10.1) and (10.2)

$$1/d \leq |A_2^{\lfloor \epsilon k \rfloor} x| \leq r/c,$$

thus by (10.4) and (10.5)

$$\begin{aligned} \rho_2(x) = |\det A_2|^{-\lfloor \epsilon k \rfloor} \rho_2(A_2^{\lfloor \epsilon k \rfloor} x) &\leq |\det A_2|^{-\lfloor \epsilon k \rfloor} \sup\{\rho_2(z) : 1/d \leq |z| \leq r/c\} \\ &\leq |\det A_1|^{-k} |\det A_2| d_2 \leq \rho_1(x) c_1^{-1} d_2 |\det A_2|. \end{aligned}$$

Similarily

$$\rho_2(x) \geq \rho_1(x) d_1^{-1} \inf\{\rho_2(z) : 1/d \leq |z| \leq r/c\} \geq \rho_1(x) d_1^{-1} c_2.$$

Since $x \in \mathbb{R}^n \setminus \{0\}$ was arbitrary the quasi-norms $\rho_1$ and $\rho_2$ are equivalent.

Conversely, assume $\rho_1$ and $\rho_2$ are equivalent, i.e., there is a constant $C > 0$ so that $1/C\rho_2(x) \leq \rho_1(x) \leq C\rho_2(x)$. For any $z \in \mathbb{R}^n$ with $|z| = 1$ we have by (10.4)

$$\begin{aligned}
(10.6) \quad \rho_1(A_1^k A_2^{-\lfloor \epsilon k \rfloor} z) &= |\det A_1|^k \rho_1(A_2^{-\lfloor \epsilon k \rfloor} z) \leq C|\det A_1|^k \rho_2(A_2^{-\lfloor \epsilon k \rfloor} z) \\
&\leq C|\det A_1|^k |\det A_2|^{-\lfloor \epsilon k \rfloor} \rho_2(z) \leq C|\det A_2| \sup\{\rho_2(x) : |x| = 1\} = D.
\end{aligned}$$

Clearly, for any $r > 0$, $\{x : \rho_1(x) \leq r\}$ is a bounded set in $\mathbb{R}^n$ if $\rho_1$ is a step homogeneous quasi-norm. By Lemma 2.4 the same is true for any quasi-norm. Hence there is a constant $d > 0$ so that $\{x : \rho_1(x) \leq D\} \subset \{x : |x| \leq d\}$. Therefore by (10.6), $|A_1^k A_2^{-\lfloor \epsilon k \rfloor} z| \leq d$. Since $z$ was arbitrary we obtain (10.2). This finishes the proof of the lemma. $\square$

Our next goal is to show the following theorem.

THEOREM 10.3. *Let $\rho_1$ and $\rho_2$ be the quasi-norms associated to dilations $A_1$ and $A_2$, respectively. Then $\rho_1$ and $\rho_2$ are equivalent if and only if for all $r > 1$ and all $m = 1, 2, \ldots$*

$$(10.7) \quad \operatorname{span} \bigcup_{|\lambda| = r^\epsilon} \ker(A_1 - \lambda Id)^m = \operatorname{span} \bigcup_{|\lambda| = r} \ker(A_2 - \lambda Id)^m,$$

*where $\epsilon$ is given in (10.3). In (10.7) we think of $A_1$ and $A_2$ as linear maps on $\mathbb{C}^n$.*

Since $A_i$ ($i = 1, 2$) is a real matrix then the complex subspace

$$\operatorname{span}(\ker(A_i - \lambda Id)^m \cup \ker(A_i - \overline{\lambda} Id)^m) \subset \mathbb{C}^n,$$

is in fact a complexification of the real subspace

$$\operatorname{span}(\ker(A_i - \lambda Id)^m \cup \ker(A_i - \overline{\lambda} Id)^m) \cap \mathbb{R}^n,$$

for any $\lambda \in \mathbb{C}$, and $m = 1, 2, \ldots$ Moreover, the conditions (10.1) and (10.2) are equivalent to respectively

$$(10.8) \quad \inf_{z \in \mathbb{C}^n \setminus \{0\}} \inf_{k \in \mathbb{Z}} |A_1^k z|/|A_2^{\lfloor \epsilon k \rfloor} z| > 0,$$

$$(10.9) \quad \sup_{z \in \mathbb{C}^n \setminus \{0\}} \sup_{k \in \mathbb{Z}} |A_1^k z|/|A_2^{\lfloor \epsilon k \rfloor} z| < \infty.$$

LEMMA 10.4. *Suppose $A$ is $n \times n$ complex matrix. For any $z \in \mathbb{C}^n$, $r > 0$, and $m = 0, 1, \ldots$*

$$(10.10) \quad z \in E(r, m+1) \setminus E(r, m), \quad \text{where}$$

$$(10.11) \quad E(r, m) = E(A; r, m) = \operatorname{span}\left( \bigcup_{|\lambda| = r} \ker(A - \lambda Id)^m \cup \bigcup_{|\lambda| < r} \ker(A - \lambda Id)^n \right),$$

*if and only if $|A^k z| \sim k^m r^k$ as $k \to \infty$, i.e., there is a constant $c > 0$ and $k_0 \in \mathbb{N}$ so that*

$$(10.12) \quad 1/c k^m r^k \leq |A^k z| \leq c k^m r^k \quad \text{for all } k \geq k_0.$$

PROOF. Suppose $J$ is a $j$ by $j$ Jordan block associated with an eigenvalue $\lambda$, i.e.,

$$(10.13) \qquad J = \begin{pmatrix} \lambda & 1 & 0 & \cdots & \cdots \\ & \lambda & 1 & 0 & \cdots \\ & & \ddots & \ddots & \ddots \\ & & & \lambda & 1 \\ & & & & \lambda \end{pmatrix}.$$

The $k$th iterate of $J$ is equal to

$$(10.14) \qquad J^k = \begin{pmatrix} p_0(k)\lambda^k & p_1(k)\lambda^{k-1} & p_2(k)\lambda^{k-2} & \cdots & p_{j-1}(k)\lambda^{k-j+1} \\ & p_0(k)\lambda^k & p_1(k)\lambda^{k-1} & p_2(k)\lambda^{k-2} & \vdots \\ & & \ddots & \ddots & \vdots \\ & & & \ddots & \vdots \\ & & & p_0(k)\lambda^k & p_1(k)\lambda^{k-1} \\ & & & & p_0(k)\lambda^k \end{pmatrix},$$

where $p_i(k)$ is given by the recursive formula

$$p_i(k) = \begin{cases} 1 & i=0, \ k=0,1,\ldots, \\ 0 & i=1,\ldots,j-1, \ k=0,\ldots,i-1, \\ p_{i-1}(k-1) + p_i(k-1) & i=1,\ldots,j-1, \ k=i,i+1,\ldots \end{cases}$$

In particular, $p_1(k) = k$, $p_2(k) = k(k-1)/2$, and by induction

$$(10.15) \qquad p_i(k) = \frac{k(k-1)\ldots(k-i+1)}{i!}.$$

Take any $0 \neq z = (z_1, \ldots, z_j) \in \mathbb{C}^j$. Let $m = 0, \ldots, j-1$ be such that

$$z \in \ker(J - \lambda Id)^{m+1} \quad \text{and} \quad z \notin \ker(J - \lambda Id)^m.$$

Equivalently, $m$ is the unique index which satisfies $z_{m+1} \neq 0$ and $z_{i+1} = 0$ for all $i > m$. If $\lambda \neq 0$ then by (10.14) and (10.15)

$$\frac{|J^k z|}{k^m |\lambda|^k} \to \frac{|z_{m+1}|}{|\lambda|^m m!} \quad \text{as } k \to \infty.$$

This shows Lemma 10.4 in the case when the matrix $A$ is a single Jordan block.

In the general case we can use the Jordan Theorem to write $A$ as $A = UBU^{-1}$, where $U$ is a nonsingular $n \times n$ matrix and $B = \bigoplus_{i=1}^p J_i$, where $J_i$ is a Jordan block of size $j_i$ associated with the eigenvalue $\lambda_i$. We can assume that $|\lambda_1| \leq \ldots \leq |\lambda_p|$ and $j_1 + \ldots + j_p = n$. For the convenience we define $l_1 = 0$, $l_i = j_1 + \ldots + j_{i-1}$ for $i = 2, \ldots, p$. We define the basis $\{v_l : l = 1, \ldots, n\}$ of Jordan decomposition by $v_l = Ue_l$, where $\{e_l : l = 1, \ldots, n\}$ denotes the standard basis in $\mathbb{C}^n$.

Note that $E(r, m) \subset E(r', m')$ if $r < r'$ or if $r = r'$ and $m \leq m'$, where $E(r, m)$ is given by (10.11). Also if $r \neq |\lambda_i|$ for all $i = 1, \ldots, p$, or $m = 0$, then

$E(r,m) = \{0\}$ if $r < |\lambda_1|$ and otherwise $E(r,m) = E(|\lambda_i|,n)$, where $i$ is the largest index so that $|\lambda_i| < r$. Since $E(|\lambda_p|,n) = \mathbb{C}^n$ therefore we can express $\mathbb{C}^n$ as

$$(10.16) \qquad \mathbb{C}^n = E(0,n) \cup \bigcup_{\substack{i=1 \\ |\lambda_i| \neq 0}}^{p} \bigcup_{m=0}^{n-1} E(|\lambda_i|, m+1) \setminus E(|\lambda_i|, m).$$

If $z \in E(0,n)$ then (10.10) can not be fulfilled for any $r > 0$ and $|A^k z| = 0$ for all $k \geq n$. Suppose that $z \in E(|\lambda_{i_0}|, m_0+1) \setminus E(|\lambda_{i_0}|, m_0)$ for some $i_0 = 1, \ldots, p$, $|\lambda_{i_0}| \neq 0$, and $m_0 = 0, \ldots, n-1$. Let $i_1 = \min\{i : |\lambda_i| = |\lambda_{i_0}|\}$, and $i_2 = \max\{i : |\lambda_i| = |\lambda_{i_0}|\}$. Observe that $E(|\lambda_{i_0}|, m)$ consists precisely of vectors $\sum_{l=1}^{n} a_l v_l$, where $a_l \in \mathbb{C}$, and $a_l = 0$ if either $l_{i_2} < l$ or $l_i + 1 + m \leq l \leq l_i + j_i$ for some $i$, $i_1 \leq i \leq i_2$. Hence, we can write $z$ as $\sum_{l=1}^{l_{i_2}} a_l v_l$. By (10.14) we have for $k \in \mathbb{N}$,

$$(10.17) \qquad \begin{aligned} A^k z &= \sum_{i=1}^{i_2} A^k \left( \sum_{l=l_i+1}^{l_i+j_i} a_l v_l \right) = \sum_{i=1}^{i_2} U \left( \sum_{l=l_i+1}^{l_i+j_i} a_l B^k e_l \right) \\ &= \sum_{i=1}^{i_2} U \left( \sum_{l=l_i+1}^{l_i+j_i} \sum_{l'=l}^{l_i+j_i} a_{l'} p_{l'-l}(k) \lambda_i^{k-(l'-l)} e_l \right) \\ &= \sum_{i=1}^{i_2} \sum_{l=l_i+1}^{l_i+j_i} \left[ \sum_{l'=l}^{l_i+j_i} a_{l'} p_{l'-l}(k) \lambda_i^{k-(l'-l)} \right] v_l. \end{aligned}$$

Since $z \in E(|\lambda_{i_0}|, m_0+1) \setminus E(|\lambda_{i_0}|, m_0)$, we have $a_l = 0$ if $l_i + 1 + m_0 + 1 \leq l \leq l_i + j_i$ for some $i$, $i_1 \leq i \leq i_2$. Also there exists $i'$, $i_1 \leq i' \leq i_2$, $m_0 + 1 \leq j_{i'}$ such that $a_{l'} \neq 0$, where $l' = l_{i'} + 1 + m_0$. Therefore, all the coefficients of the $v_l$'s in the brackets (10.17) are dominated asymptotically by $k^{m_0} |\lambda_0|^k$ as $k \to \infty$ and at least one coefficient (the coefficient of $v_{l_{i'}+1}$) behaves asymptotically as $k^{m_0} |\lambda_0|^k$ as $k \to \infty$ by (10.15). Since the norm $|U \cdot|$ is equivalent to the standard norm $|\cdot|$, this shows that there exists a constant $c > 0$, and $k_0 \in \mathbb{N}$ so that

$$1/c k^{m_0} |\lambda_0|^k \leq |A^k z| \leq c k^{m_0} |\lambda_0|^k \qquad \text{for all } k \geq k_0,$$

i.e., (10.12) holds.

This combined with (10.16) also shows the converse implication. $\square$

PROOF OF THEOREM 10.3. Lemma 10.2 says that the quasi-norms $\rho_1$ and $\rho_2$ are equivalent if and only if (10.8) and/or (10.9) hold. Assume that this happens. For $r \geq 0$ and $m = 0, 1, \ldots$, let $E(A_i; r, m)$ be the linear space given by (10.11) corresponding to the dilation $A_i$, $i = 1, 2$. By Lemma 10.4, if $z \in E(A_1; r, m+1) \setminus E(A_1; r, m)$ then $|A_1^k z| \sim k^m r^k$ as $k \to \infty$ and by (10.8) and (10.9), $|A_2^{\lfloor \epsilon k \rfloor} z| \sim k^m r^k$, so $|A_2^k z| \sim k^m r^{k/\epsilon}$ as $k \to \infty$. Therefore, $z \in E(A_2; r^{1/\epsilon}, m+1) \setminus E(A_2; r^{1/\epsilon}, m)$. Since the sets $E(A_i; r, m+1) \setminus E(A_i; r, m)$, $r > 1$, $m = 0, 1, \ldots$ partition $\mathbb{C}^n \setminus \{0\}$ for $i = 1, 2$, we have $E(A_1; r^\epsilon, m) = E(A_2; r, m)$ for all $r > 0$, $m = 0, 1, \ldots$. Analogously, by considering matrices $A_1^{-1}$ and $A_2^{-1}$ we have $E(A_1^{-1}; r^\epsilon, m) = E(A_2^{-1}; r, m)$ for all $r > 0$, $m = 0, 1, \ldots$. Since $E(A_i; r, m) \cap E(A_i^{-1}; r^{-1}, m) = \mathrm{span} \bigcup_{|\lambda|=r} \ker(A_i - \lambda Id)^m$ for all $r > 0$, $m = 0, 1, \ldots$, we have (10.7). By reversing this argument we obtain the converse implication. $\square$

We are ready to state the classification theorem for dilations generating the same anisotropic Hardy space $H^p$, $0 < p \leq 1$.

THEOREM 10.5. *Suppose we have two dilations $A_1$ and $A_2$ on $\mathbb{R}^n$. The following are equivalent:*

*(i) the quasi-norms $\rho_1$ and $\rho_2$ associated to $A_1$ and $A_2$, respectively, are equivalent,*
*(ii) (10.7) holds for all $r > 1$, $m = 1, 2, \ldots,$*
*(iii) the anisotropic Hardy spaces $H^p$ associated to $A_1$ and $A_2$ are the same for some $0 < p \leq 1$,*
*(iv) the anisotropic Hardy spaces $H^p$ associated to $A_1$ and $A_2$ are the same for all $0 < p \leq 1$.*

REMARK. We claim that the atoms introduced in Definition 4.1 can be alternatively defined as follows. Suppose $A$ is a dilation and $\rho$ its associated quasi-norm. We say a triplet $(p, q, s)$ is *admissible* (with respect to dilation $A$) if $0 < p \leq 1$, $1 \leq q \leq \infty$, $p < q$, $s \in \mathbb{N}$, and $s \geq \lfloor (1/p - 1) \ln b / \ln \lambda_- \rfloor$, where $b = |\det A|$, and $\lambda_-$ is such that (2.1) holds. A $(p, q, s)$-atom (associated with dilation $A$) is a function $a$ such that

(10.18)
$$\operatorname{supp} a \subset \{x : \rho(x_0 - x) \leq r\} \quad \text{for some } r > 0, x_0 \in \mathbb{R}^n,$$

(10.19)
$$\|a\|_q \leq r^{1/q - 1/p},$$

(10.20)
$$\int_{\mathbb{R}^n} a(x) x^\alpha dx = 0 \quad \text{for } |\alpha| \leq s.$$

Indeed, if $\rho$ is the step homogeneous quasi-norm then the above conditions for $r = b^j$, $j \in \mathbb{Z}$ coincide with Definition 4.1. If $\rho$ is a general quasi-norm then by Lemma 2.4 there is a constant $c > 0$ so that for every atom $a$ satisfying (10.18), (10.19), and (10.20), $ca$ is an atom in the sense of Definition 4.1. And vice versa. Therefore, the definition of atoms is independent of the choice of a quasi-norm up to the equivalence of a multiplicative constant.

We can also replace conditions (10.18) and (10.19) by

(10.21)
$$\operatorname{supp} a \subset x_0 + A^j \mathbf{B}(0, 1) \quad \text{for some } j \in \mathbb{Z}, x_0 \in \mathbb{R}^n,$$

(10.22)
$$\|a\|_q \leq |A^j \mathbf{B}(0, 1)|^{1/q - 1/p},$$

where $\mathbf{B}(0, 1) = \{x : |x| < 1\}$.

PROOF. Theorem 10.3 states that (i) $\iff$ (ii). (i) $\implies$ (iv) is a consequence of the above Remark. (iv) $\implies$ (iii) is automatic. It suffices to show (iii) $\implies$ (ii).

Assume that (iii) holds for some $0 < p \leq 1$. Our goal is to show (ii). Denote the Hardy space associated to the $n \times n$ dilation matrix $A$ by $H_A^p(\mathbb{R}^n)$ or simply $H_A^p$. We claim that $H_{A_1}^p = H_{A_2}^p$ implies that there is a constant $C > 0$ so that

(10.23) $\qquad 1/C \|f\|_{H_{A_1}^p} \leq \|f\|_{H_{A_2}^p} \leq C \|f\|_{H_{A_1}^p} \qquad \text{for all } f \in H_{A_1}^p = H_{A_2}^p.$

By Theorem 6.9 and the following Remark, (10.23) is equivalent to the same condition being satisfied for all bounded, compactly supported functions $f$ with vanishing moments up to order $s$, where $s$ is large enough so that triplet $(p, \infty, s)$ is admissible for both $A_1$ and $A_2$. The family of such functions consists precisely of scalar multiples of $(p, \infty, s)$-atoms (associated with $A_1$ or $A_2$). It is not hard to see that the equivalence of norms for atoms implies the corresponding statement for all elements of $H_{A_1}^p = H_{A_2}^p$, i.e., (10.23).

If (10.23) fails then we could find a sequence of elements $(f_i)_{i\in\mathbb{N}}$ in $H^p_{A_1}$ so that $||f_i||_{H^p_{A_1}} \leq 2^{-i}$ and $||f_i||_{H^p_{A_2}} = 1$ for all $i \in \mathbb{N}$ (or the similar statement with the norms $||\cdot||_{H^p_{A_1}}$, $||\cdot||_{H^p_{A_2}}$ interchanged). For any choice of vectors $k_i \in \mathbb{R}^n$ the series $\sum_{i\in\mathbb{N}} \tau_{k_i} f_i$ converges in $H^p_{A_1}$ (and hence in $\mathcal{S}'$), since $\sum_{i\in\mathbb{N}} ||f_i||^p_{H^p_{A_1}} < \infty$, see Proposition 4.5. Here $\tau_k f(x) = f(x-k)$ denotes the translation of $f$ by the vector $k \in \mathbb{R}^n$. We claim that for some choice of $k_i$'s, $\sum_{i\in\mathbb{N}} \tau_{k_i} f_i$ does not belong to $H^p_{A_2}$. Indeed, if $M$ denotes the (grand) maximal function associated to $A_2$ then find the sequence of numbers $(r_i)_{i\in\mathbb{N}}$ so that

$$(10.24) \qquad \int_{\mathbf{B}(0,r_i)} Mf_i(x)^p dx > 1 - 2^{-i-1} \qquad \text{for all } i \in \mathbb{N}.$$

Choose $k_i$'s so that the balls $\mathbf{B}(k_i, r_i)$ are mutually disjoint. Let $f = \sum_{i\in\mathbb{N}} \tau_{k_i} f_i \in \mathcal{S}'$. For each $j \in \mathbb{N}$, $\tau_{k_j} f_j = f - \sum_{i\neq j} \tau_{k_i} f_i$ with convergence in $\mathcal{S}'$, hence

$$M(\tau_{k_j} f_j)(x) \leq Mf(x) + \sum_{i\neq j} M(\tau_{k_i} f_i)(x) \leq \left(Mf(x)^p + \sum_{i\neq j} M(\tau_{k_i} f_i)(x)^p\right)^{1/p},$$

for all $x \in \mathbb{R}^n$. In particular

$$Mf(x)^p \geq M(\tau_{k_j} f_j)(x)^p - \sum_{i\neq j} M(\tau_{k_i} f_i)(x)^p \qquad \text{for } x \in \mathbf{B}(k_j, r_j).$$

Integrating the above and using $\tau_k M = M\tau_k$ we have

$$\int_{\mathbf{B}(k_j,r_j)} Mf(x)^p dx \geq \int_{\mathbf{B}(k_j,r_j)} \left(M(\tau_{k_j}f_j)(x)^p - \sum_{i\neq j} M(\tau_{k_i}f_i)(x)^p\right) dx$$

$$= \int_{\mathbf{B}(0,r_j)} Mf_j(x)^p dx - \sum_{i\neq j} \int_{\mathbf{B}(k_j-k_i, r_j)} Mf_i(x)^p dx$$

$$\geq 1 - 2^{-j-1} - \sum_{i\neq j} \int_{\mathbf{B}(0,r_i)^c} Mf_i(x)^p dx \geq 3/4 - \sum_{i\in\mathbb{N}} 2^{-i-1} = 1/4,$$

by (10.24) and $\int_{\mathbb{R}^n} Mf_i(x)^p dx = 1$. Summing the above over $j \in \mathbb{N}$ we have

$$\int_{\mathbb{R}^n} Mf(x)^p dx \geq \sum_{j\in\mathbb{N}} \int_{\mathbf{B}(k_j,r_j)} Mf(x)^p dx = \infty.$$

Hence, $f \notin H^p_{A_2}$ which is a contradiction of (iii). Therefore (10.23) holds.

We remark that if $A$ is a dilation then $D_A f(x) = |\det A|^{1/p} f(Ax)$ is an isometry on $H^p_A$. More precisely, if $f \in \mathcal{S}'$ and $\varphi \in \mathcal{S}$ then we define $D_A f$ by

$$\langle D_A f, \varphi \rangle = |\det A|^{1/p-1} \langle f, \varphi(A^{-1} \cdot) \rangle.$$

Indeed, by a simple calculation we have

$$M^0_\varphi f(Ax) = |\det A|^{-1/p} M^0_\varphi (D_A f)(x) \qquad \text{for } x \in \mathbb{R}^n,$$

and by the definition of the grand maximal function and a change of variables

$$(10.25) \qquad ||f||_{H^p_A} = ||D_A f||_{H^p_A} \qquad \text{for all } f \in H^p_A.$$

Consider the family of functions $f$ on $\mathbb{R}^n$ so that

(10.26) $\qquad \operatorname{supp} f \subset x_0 + A_1^{j_1} A_2^{j_2} \mathbf{B}(0,1) \qquad$ for some $j_1, j_2 \in \mathbb{Z}, x_0 \in \mathbb{R}^n$,

(10.27) $\qquad \|f\|_\infty \leq |\det A_1|^{-j_1/p} |\det A_2|^{-j_2/p},$

(10.28) $\qquad \int_{\mathbb{R}^n} f(x) x^\alpha dx = 0 \qquad$ for all $|\alpha| \leq s.$

We claim that there is a constant $C' > 0$ so that $\|f\|_{H_{A_1}^p} \leq C'$ for every $f$ satisfying (10.26)–(10.28). Indeed, for any such $f$, $D_{A_2^{j_2}} D_{A_1^{j_1}} f$ has support contained in $A_2^{-j_2} A_1^{-j_1} x_0 + \mathbf{B}(0,1)$, has $L^\infty$ norm less than 1, and has vanishing moments up to order $s$, hence is a constant multiple of an atom (satisfying (10.20)–(10.22) with $j = 0$ and $q = \infty$). The claim now follows from (10.23) and the iterative form of (10.25). Finally, let $f_0$ be a fixed function satisfying (10.26)–(10.28) with $x_0 = 0$, $j_1 = j_2 = 0$,

(10.29) $\qquad f_0(x) = \begin{cases} \delta_0 & \text{for all } x \in \mathbf{B}(3/4 e_1, 1/4), \\ 0 & \text{for all } x \notin \mathbf{B}(0, 1/2) \cup \mathbf{B}(3/4 e_1, 1/4). \end{cases}$

It is clear that if $\delta_0 > 0$ is sufficiently small a function $f_0$ satisfying the above constraints exists.

To finish the proof, assume on the contrary that (ii) fails. By (10.2) this means that either

$$\limsup_{k \to \infty} \|A_1^k A_2^{-\lfloor \epsilon k \rfloor}\| = \infty \quad \text{or} \quad \limsup_{k \to -\infty} \|A_1^k A_2^{-\lfloor \epsilon k \rfloor}\| = \infty.$$

It is not hard to see the lim sup can be replaced by lim using the idea in the proof of Theorem 10.3. For any $k \in \mathbb{Z}$ define $d(k)$ as the smallest integer so that

$$\|A_1^k A_2^{-\lfloor \epsilon k \rfloor - d(k)}\| \leq 1.$$

Clearly, we have

$$\|A_1^k A_2^{-\lfloor \epsilon k \rfloor - d(k)}\| = \|A_1^k A_2^{-\lfloor \epsilon k \rfloor - d(k)}\| \|A_2\| \|A_2\|^{-1} \geq \|A_2\|^{-1}.$$

We either have $d(k) \to \infty$ as $k \to \infty$ or $d(k) \to \infty$ as $k \to -\infty$. We will fix our attention on the case when $d(k) \to \infty$ as $k \to \infty$; the other case is identical.

For simplicity denote $Q_k = A_1^k A_2^{-\lfloor \epsilon k \rfloor - d(k)}$. Let $z_k \in \mathbb{R}^n$, $|z_k| = 1$, be such that $|Q_k z_k| = \|Q_k\| =: c(k)$. We know that $\|A_2\|^{-1} \leq c(k) \leq 1$. Let $U_k$ be a unitary matrix such that $U_k e_1 = z_k$. Consider function

(10.30) $\qquad f_k = D_{U_k^{-1} Q_k^{-1}} f_0 = D_{Q_k^{-1}} D_{U_k^{-1}} f_0,$

The function $f_k$ clearly satisfies (10.26)–(10.28) with $x_0 = 0$, $j_1 = k$, $j_2 = -\lfloor \epsilon k \rfloor - d(k)$. Since $Q_k U_k \mathbf{B}(0, 1/2) \subset \mathbf{B}(0, c(k)/2)$ and $Q_k U_k \mathbf{B}(3/4 e_1, 1/4) = Q_k \mathbf{B}(3/4 z_k, 1/4)$ then by (10.29) and (10.30)

(10.31) $\qquad \begin{aligned} & x \in \mathbf{B}(0, c(k)/2)^c \quad \text{and} \quad f_k(x) \neq 0 \implies \\ & x \in Q_k \mathbf{B}(3/4 z_k, 1/4) \quad \text{and} \quad f_k(x) = \delta_k, \end{aligned}$

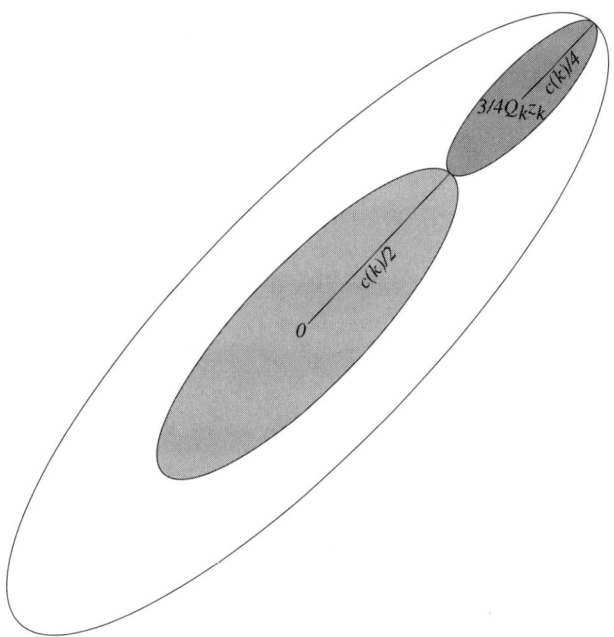

FIGURE 3. THE SUPPORT OF $f_k$ IS CONTAINED IN THE SHADED REGIONS. THE FUNCTION $f_k$ IS EQUAL TO $\delta_k$ IN THE DARKER ELLIPSE.

where $\delta_k = \delta_0 |\det A_1|^{-k/p} |\det A_2|^{\lfloor \epsilon k \rfloor/p + d(k)/p}$. Let $\varphi \in \mathcal{S}$ be a nonnegative function such that

(10.32) $$\varphi(x) = \begin{cases} 1 & \text{for } x \in \mathbf{B}(0, 1/8\|A_2\|^{-1}), \\ 0 & \text{for } x \notin \mathbf{B}(0, 3/16\|A_2\|^{-1}). \end{cases}$$

We claim that $\|M_\varphi^0 f_k\|_p \to \infty$ as $k \to \infty$. Indeed, if $z \in \mathbf{B}(3/4 Q_k z_k, 1/16\|A_2\|^{-1})$ then

(10.33)
$$\begin{aligned} M_\varphi^0 f_k(z) \geq |f_k * \varphi(z)| &= \left| \int_{\mathbb{R}^n} f_k(x) \varphi(z-x) dx \right| \\ &= \delta_k \int_{\mathbb{R}^n} \mathbf{1}_{Q_k \mathbf{B}(3/4 z_k, 1/4)}(x) \varphi(z-x) dx \\ &\geq \delta_k |\mathbf{B}(z, 1/8\|A_2\|^{-1}) \cap Q_k \mathbf{B}(3/4 z_k, 1/4)|, \end{aligned}$$

by (10.31) and (10.32) since $\varphi(z-x) \neq 0$ implies

$$x \in \mathbf{B}(0, 3/16\|A_2\|^{-1}) + \mathbf{B}(3/4 Q_k z_k, 1/16\|A_2\|^{-1}) = \mathbf{B}(3/4 Q_k z_k, 1/4\|A_2\|^{-1}),$$

and hence $x \in \mathbf{B}(0, c(k)/2)^c$. By (10.33) and Lemma 10.6 applied to $P = 1/4 Q_k$ and $r = (2c(k)\|A_2\|)^{-1}$

$$\begin{aligned} M_\varphi^0 f_k(z) &\geq \delta_k |\mathbf{B}(z - 3/4 Q_k z_k, 1/8\|A_2\|^{-1}) \cap P\mathbf{B}(0, 1)| \geq \delta_k (r/2)^n |P\mathbf{B}(0, 1)| \\ &\geq |\det Q_k| |\mathbf{B}(0, 1/4)| \delta_k (4\|A_2\|)^{-n} \\ &\geq \delta_0 |\det A_2|^{(d(k)-1)(1/p-1)} |\mathbf{B}(0, 1/4)| (4\|A_2\|)^{-n} = c |\det A_2|^{d(k)(1/p-1)}, \end{aligned}$$

by (10.4). Therefore, if $p < 1$ then

$$\int_{\mathbb{R}^n} M_\varphi^0 f_k(z)^p dz \geq \int_{\mathbf{B}(3/4Q_k z_k, 1/16\|A_2\|^{-1})} M_\varphi^0 f_k(z)^p dz$$
$$\geq c^p |\mathbf{B}(0, 1/16\|A_2\|^{-1})| |\det A_2|^{d(k)(1-p)} \to \infty \quad \text{as } k \to \infty.$$

By Theorem 7.1, $\|f_k\|_{H^p_{A_1}} \to \infty$ as $k \to \infty$ which contradicts $\|f_k\|_{H^p_{A_1}} \leq C'$. Therefore, (ii) must necessarily hold if $p < 1$.

The case $p = 1$ requires a special argument. We have $\|f_k\|_1 = \|f_0\|_1$ and $\operatorname{supp} f_k \subset Q_k \mathbf{B}(0,1)$ for all $k \in \mathbb{N}$. Since $\mathcal{S}$ is a separable space there is a subsequence $(f_{k_i})_{i \in \mathbb{N}}$ converging to some $f \in \mathcal{S}'$, and $r = \lim_{i \to \infty} c(k_i)$ exists. Clearly $f$ is a regular Borel measure with compact support which is singular with respect to the Lebesgue measure since $|\operatorname{supp} f_k| \to 0$ as $k \to \infty$. Furthermore, $f \neq 0$ which can be verified by testing $f$ with an appropriate (nonnegative, radial) test function $\varphi \in \mathcal{S}$ vanishing on $\mathbf{B}(0, r/2)$. Let $M$ denotes the (grand) maximal function associated to $A_1$. By Fatou's Lemma

$$\int_{\mathbb{R}^n} Mf(x) dx \leq \int_{\mathbb{R}^n} \liminf_{i \to \infty} Mf_{k_i}(x) dx \leq \liminf_{i \to \infty} \int_{\mathbb{R}^n} Mf_{k_i}(x) dx$$
$$\leq \liminf_{i \to \infty} \|f_{k_i}\|_{H^1_{A_1}} \leq C'.$$

Hence $f \in H^1_{A_1}$. But $f \notin L^1$ which is a contradiction of $H^1_{A_1}(\mathbb{R}^n) \subset L^1(\mathbb{R}^n)$. This finishes the proof of Theorem 10.5. $\square$

LEMMA 10.6. *Suppose* $\Gamma = P\mathbf{B}(0,1) = \{x \in \mathbb{R}^n : |P^{-1}x| < 1\}$ *is an ellipsoid, where $P$ is some nondegenerate $n \times n$ matrix. For any $0 < r \leq 1/2$ we have*

$$\frac{|\mathbf{B}(z, \|P\|r) \cap \Gamma|}{|\Gamma|} \geq (r/2)^n \qquad \text{for all } z \in \mathbf{B}(0, \|P\|r/2).$$

PROOF. Let $c = \|P\|$. For any $z \in \mathbf{B}(0, cr/2)$, $\mathbf{B}(0, cr/2) \subset \mathbf{B}(z, cr)$ hence

$$\frac{|\mathbf{B}(z, cr) \cap \Gamma|}{|\Gamma|} \geq \frac{|\mathbf{B}(0, cr/2) \cap P\mathbf{B}(0,1)|}{|P\mathbf{B}(0,1)|} = \frac{|P^{-1}\mathbf{B}(0, cr/2) \cap \mathbf{B}(0,1)|}{|\mathbf{B}(0,1)|}$$
$$\geq \frac{|\mathbf{B}(0, r/2) \cap \mathbf{B}(0,1)|}{|\mathbf{B}(0,1)|} = (r/2)^n. \qquad \square$$

We close this section with a simple example of a distribution that differentiates most anisotropic $H^p$ spaces. Although this example gives only a necessary condition for $H^p_{A_1} = H^p_{A_2}$, $0 < p < 1$, it motivated the author toward Theorem 10.5.

LEMMA 10.7. *Suppose $0 < p < 1$ and $A$ is a dilation. For fixed $z \in \mathbb{R}^n \setminus \{0\}$, let $f = \delta_0 - \delta_z \in \mathcal{S}'$, where $\delta_z$ denotes the point mass at $z$. Then*

(10.34) $$\int_{\mathbb{R}^n} M_\varphi^0 f(x)^p dx < \infty \iff \sum_{l=1}^{\infty} |A^{-l}z|^p |\det A|^{(1-p)l} < \infty,$$

*where $\varphi \in \mathcal{S}$, $\int \varphi \neq 0$.*

PROOF. Since $M_\varphi^0 \delta_0(x) \approx 1/\rho(x)$, where $\rho$ is a quasi-norm associated to $A$, $M_\varphi^0 \delta_0$ is locally in $L^p$ ($0 < p < 1$). Therefore, only the behavior of $M_\varphi^0$ at infinity is of importance. For simplicity we choose a nonnegative $\varphi \in \mathcal{S}$ so that $\varphi(x) = 1$ for all $x \in B_0^{\bullet}$ and $\operatorname{supp}\varphi \subset B_1$, where $B_i$'s come from Definition 2.5. By the Mean Value Theorem

$$\int_{B_1 \setminus B_0} |\langle \nabla\varphi(x), w\rangle|^p dx > 0 \qquad \text{for any } w \in \mathbb{R}^n \setminus \{0\},$$

where $\nabla\varphi$ is the gradient of $\varphi$. By continuity

$$\inf_{|w|=1} \int_{B_1 \setminus B_0} |\langle \nabla\varphi(x), w\rangle|^p dx > 0.$$

Since $|\varphi(x) - \varphi(x-w) - \langle \nabla\varphi(x), w\rangle|/|w| \to 0$ as $|w| \to 0$ uniformly over $x \in B_1 \setminus B_0$, there is a constant $c > 0$ so that

(10.35) $$\int_{B_1 \setminus B_0} |\varphi(x) - \varphi(x-w)|^p dx \geq c|w|^p \qquad \text{for all } |w| \leq 1.$$

We have

(10.36)
$$\begin{aligned} M_\varphi^0 f(x) &= \sup_{k \in \mathbb{Z}} b^{-k} \left| \int f(y)\varphi(A^{-k}(x-y))dy \right| \\ &= \sup_{k \in \mathbb{Z}} b^{-k} |\varphi(A^{-k}x) - \varphi(A^{-k}(x-z))|, \end{aligned}$$

where $b = |\det A|$. Without loss of generality we can assume that $|z| \leq 1$. Suppose $x \in B_{l+1} \setminus B_l$ for some $l \geq l_0$, where $l_0$ is sufficiently large so that $(B_l)^{\mathbf{c}} + \mathbf{B}(0,1) \subset (B_{l-1})^{\mathbf{c}}$ for all $l \geq l_0$. Since $x - z \in (B_{l-1})^{\mathbf{c}}$ the supremum in (10.36) runs effectively only for $k \geq l - 1$, and

$$M_\varphi^0 f(x) \leq b^{1-l} \sup_{y \in \mathbb{R}^n} |\nabla\varphi(y)| \sup_{k \geq l-1} |A^{-k}z| \leq C b^{-l} |A^{-l}z|.$$

Therefore

$$\int_{B_{l+1} \setminus B_l} M_\varphi^0 f(x)^p dx \leq C^p b^{l+1} b^{-lp} |A^{-l}z|^p = C^p b^{(1-p)l+1} |A^{-l}z|^p.$$

On the other hand by (10.35)

$$\begin{aligned} \int_{B_{l+1} \setminus B_l} M_\varphi^0 f(x)^p dx &\geq b^{-lp} \int_{B_{l+1} \setminus B_l} |\varphi(A^{-l}x) - \varphi(A^{-l}(x-z))|^p dx \\ &= b^{-lp} b^l \int_{B_1 \setminus B_0} |\varphi(x) - \varphi(x - A^{-l}z)|^p dx \geq c b^{(1-p)l} |A^{-l}z|^p. \end{aligned}$$

Summing the last two estimates over $l \geq l_0$ yields (10.34). $\square$

COROLLARY 10.8. *Suppose $0 < p < 1$, $A$ is a dilation, and $z \in \mathbb{R}^n \setminus \{0\}$. The following are equivalent:*

$$\delta_0 - \delta_z \in H_A^p,$$
$$|A^{-l}z||\det A|^{(1/p-1)l} \to 0 \quad \text{as } l \to \infty,$$
$$z \in \operatorname{span}\left( \bigcup_{|\lambda| > |\det A|^{1/p-1}} \ker(A - \lambda Id)^n \right).$$

PROOF. Corollary follows easily from Lemmas 10.4 and 10.7. □

EXAMPLE. Suppose $A_1$ and $A_2$ are dilations. Let $\lambda_1^i, \ldots, \lambda_n^i$ be eigenvalues of $A_i$ (taken according to multiplicity) so that $|\lambda_1^i| \leq \ldots \leq |\lambda_n^i|$, $i = 1, 2$. Suppose that $H_{A_1}^p = H_{A_2}^p$ for all $0 < p < 1$. By Corollary 10.8 this implies $(r = |\det A_1|^{1/p-1})$

$$\operatorname{span} \bigcup_{|\lambda| > r} \ker(A_1 - \lambda Id)^n = \operatorname{span} \bigcup_{|\lambda| > r^{1/\epsilon}} \ker(A_2 - \lambda Id)^n \qquad \text{for all } r > 1,$$

where $\epsilon$ is given by (10.3). In particular, by counting dimensions we have

(10.37) $$|\lambda_j^1|^{1/\ln|\det A_1|} = |\lambda_j^2|^{1/\ln|\det A_2|} \qquad \text{for all } j = 1, \ldots, n.$$

Naturally this condition falls short of sufficiency (in comparison with (10.7), for example). Nevertheless, it does give a quick way of checking whether two dilations might generate the same anisotropic $H^p$ spaces. In fact, (10.37) comes very close to characterizing those Hardy spaces which are equivalent up to a linear transformation, as we will see in Theorem 10.10.

DEFINITION 10.9. We say that $H_{A_1}^p$ and $H_{A_2}^p$ are equivalent up to a linear transformation, and write $H_{A_1}^p \cong H_{A_2}^p$ if there is a nonsingular $n \times n$ matrix $P$ such that $D_P$ is an isomorphism between $H_{A_1}^p$ and $H_{A_2}^p$. We say that two quasi-norms are equivalent up to a linear transformation if there is a constant $c > 0$ and a nonsingular $n \times n$ matrix $P$ such that

$$1/c \rho_1(x) \leq \rho_2(Px) \leq c \rho_1(x) \qquad \text{for all } x \in \mathbb{R}^n.$$

THEOREM 10.10. Suppose we have two dilations $A_1$ and $A_2$ on $\mathbb{R}^n$. The following are equivalent:
(i) the quasi-norms $\rho_1$ and $\rho_2$ associated to $A_1$ and $A_2$, respectively, are equivalent up to a linear transformation,
(ii) for all $r > 1$ and $m = 1, 2, \ldots$ we have

(10.38) $$\sum_{|\lambda| = r^\epsilon} \dim \ker(A_1 - \lambda Id)^m = \sum_{|\lambda| = r} \dim \ker(A_2 - \lambda Id)^m,$$

where $\epsilon$ is given by (10.3),
(iii) $H_{A_1}^p \cong H_{A_2}^p$ for all $0 < p \leq 1$,
(iv) $H_{A_1}^p \cong H_{A_2}^p$ for some $0 < p \leq 1$.

PROOF. Without loss of generality we can assume that the dilations $A_1$ and $A_2$ have only positive eigenvalues and $\det A_1 = \det A_2 = b$ by Theorem 10.5.

Assume (i) holds. Note that $\rho_2(P \cdot)$ is a quasi-norm associated with the dilation $P^{-1} A_2 P$. Indeed,

$$\rho_2(P(P^{-1} A_2 P x)) = |\det A_2| \rho_2(Px) = |\det(P^{-1} A_2 P)| \rho_2(x).$$

Since the quasi-norms $\rho_1$ and $\rho_2(P \cdot)$ are equivalent

$$\ker(A_1 - rId)^m = \ker(P^{-1} A_2 P - rId)^m = \ker(P^{-1}(A_2 - rId)^m P)$$
$$= P^{-1}(\ker(A_2 - rId)^m),$$

for any $r > 1$, $m = 1, 2, \ldots$ by Theorem 10.5. Hence (ii) holds.

## 10. CLASSIFICATION OF DILATIONS

Assume (ii) holds, i.e.,

(10.39) $\quad \dim \ker(A_1 - rId)^m = \dim \ker(A_2 - rId)^m, \qquad$ for all $r > 1, m = 1, 2, \ldots$

If $r$ is an eigenvalue of $A_1$ then the number of Jordan blocks of size $\geq m$ corresponding to $r$ is equal to $\dim \ker(A_1 - rId)^m - \dim \ker(A_1 - rId)^{m-1}$. If $r$ is not an eigenvalue of $A_1$ this number equals 0 regardless of $m$. By (10.39) the number of Jordan blocks of size $m$ corresponding to the eigenvalue $r$ is the same for both $A_1$ and $A_2$ and therefore the matrices $A_1$ and $A_2$ have equivalent Jordan decompositions. So there is a nonsingular $n \times n$ matrix $P$ such that $A_1 = P^{-1}A_2P$. Choose any $\varphi \in \mathcal{S}$ with $\int \varphi \neq 0$. Let $\varphi_k(x) = b^{-k}\varphi(A_1^{-k}x)$ and $\psi_k(x) = b^{-k}\psi(A_2^{-k}x)$, where $\psi(x) = |\det P^{-1}|\varphi(P^{-1}x)$. Given any $f \in \mathcal{S}'$ we have

$$(f * \varphi_k)(x) = b^{-k} \int f(y)\varphi(P^{-1}A_2^{-k}P(x-y))dy$$

$$= b^{-k}|\det P^{-1}| \int f(P^{-1}y)\varphi(P^{-1}A_2^{-k}(Px - y))dy$$

$$= \int f(P^{-1}y)\psi_k(Px - y)dy = |\det P|^{1/p}(D_{P^{-1}}f * \psi_k)(Px).$$

Therefore, $\|M_\varphi^0 f\|_p = \|M_\psi^0 D_{P^{-1}} f\|_p$, where the maximal functions are associated to $A_1$ and $A_2$, respectively. Thus, $\|f\|_{H_{A_1}^p} \sim \|D_{P^{-1}}f\|_{H_{A_2}^p}$ for any $f$, and $H_{A_1}^p \cong H_{A_2}^p$ for any $0 < p \leq 1$, thus (iii) holds. Since (iii) trivially implies (iv), it suffices to show (iv) $\implies$ (i).

Assume (iv) holds, i.e., for some $0 < p \leq 1$, $\|f\|_{H_{A_1}^p} \sim \|D_{P^{-1}}f\|_{H_{A_2}^p}$ for any $f \in H_{A_1}^p$. Consider the family of functions $f$ on $\mathbb{R}^n$ such that

(10.40) $\quad \operatorname{supp} f \subset x_0 + A_1^{j_1} P^{-1} A_2^{j_2} P \mathbf{B}(0,1) \qquad$ for some $j_1, j_2 \in \mathbb{Z}, x_0 \in \mathbb{R}^n$,

(10.41) $\quad \|f\|_\infty \leq b^{-j_1/p - j_2/p}$,

(10.42) $\quad \int_{\mathbb{R}^n} f(x) x^\alpha dx = 0 \qquad$ for all $|\alpha| \leq s$.

We claim that there is a constant $C' > 0$ so that $\|f\|_{H_{A_1}^p} \leq C'$ for every $f$ satisfying (10.40)–(10.42). Indeed, for any such $f$, $D_{A_2^{j_2}} D_{P^{-1}} D_{A_1^{j_1}} f$ has support contained in $A_2^{-j_2} P A_1^{-j_1} x_0 + P\mathbf{B}(0,1)$, has $L^\infty$ norm less than $|\det P|^{-1/p}$, and has vanishing moments up to order $s$, hence is a constant multiple of an atom. The claim now follows from the hypothesis and the iterative form of (10.25). By the proof of Theorem 10.5 we conclude that the dilations $A_1$ and $P^{-1}A_2P$ have equivalent quasi-norms $\rho_1$ and $\tilde{\rho}_2$, respectively. Define $\rho_2(x) = \tilde{\rho}_2(P^{-1}x)$. Clearly $\rho_2$ is a quasi-norm associated with the dilation $A_2$ and $\rho_1$ and $\rho_2(P\cdot)$ are equivalent. This shows (i) and ends the proof. $\square$

CHAPTER 2

# Wavelets

## 1. Introduction

In this chapter we present constructions of orthogonal and tight frame wavelets in the Schwartz class. We investigate the limitations on regularity of orthogonal wavelets imposed by the form of a dilation. We show that sufficiently regular wavelets form an unconditional basis for the anisotropic Hardy space associated this dilation.

**Historical background.** The theory of wavelets is a relatively new area of mathematics. The first example of an object now called a wavelet is the Haar function introduced by Haar [Ha] in 1910. Haar showed that the appropriate translates and dilates of Haar function form an orthonormal basis of $L^2([0,1])$. The second example of wavelets has been introduced by Strömberg [Sö] in 1981. Strömberg wavelets can be constructed to be $C^r$ smooth with exponential decay at infinity for any integer $r \geq 1$. Strömberg has also shown that they form an unconditional basis for the Hardy space $H^p(\mathbb{R}^n)$, $0 < p \leq 1$.

The theory of wavelets has taken off with the construction of wavelets in the Schwartz class by Meyer [Me] in 1985. Soon after Daubechies [Da1, Da2] has constructed compactly supported wavelets in $C^r$ class for any integer $r \geq 1$. In the first few years of its existence the theory of wavelets has grown exponentially and it is still a very active area of research both in pure and applied mathematics. For example, wavelets turn out to have many advantages in studying various function spaces. They form an unconditional bases for a variety of function spaces, e.g. $L^p$, Hardy, Hölder, Sobolov, and Besov spaces. In the applied sciences wavelets are successfully used in many areas of signal analysis, e.g. image compression, noise reduction, feature extraction, etc., see [JMR]. Here we are going to discuss only the part of the theory concerned with wavelets in Euclidean spaces.

**Description of the chapter.** Dai, Larson and Speegle [DL, DLS] have shown the existence of orthogonal wavelets (minimally supported in frequency) for all dilations. However, the construction of more regular wavelets is a complex process which is apparently impossible for "most" of dilations, see Theorem 3.1.

Our initial goal is the construction of regular orthogonal wavelets for a general dilation matrix $A$ preserving some lattice $\Gamma$. Strichartz [Sr] has constructed $r$-regular wavelets with an associated $r$-regular multiresolution analysis for every $r$ and for a wide class of dilations having a Haar type wavelet basis, or equivalently a self-affine tiling, see [GM]. This result was extended to all dilations preserving some lattice in [Bo4].

However, the problem of finding $\infty$-regular multiwavelets, i.e., in the Schwartz class, for a general dilation $A$ preserving some lattice is still open. Nevertheless,

# 1. INTRODUCTION

we are able to show the existence of orthogonal wavelets in the Schwartz class associated with some positive power of $A$. This is the content of Section 2.

In Section 3 we show that if move beyond the class of dilations preserving some lattice, regular wavelets may not exists. In fact, we show that for a large class of dilations all multiwavelets must be combined minimally supported in frequency. Nevertheless, in Section 4 we show that $\infty$-regular tight frame wavelets exist for any dilation matrix.

In Section 5 we show that $r$-regular wavelets associated with a dilation $A$ form an unconditional basis for the anisotropic Hardy space $H_A^p$ associated with $A$. This generalizes the result of Meyer [Me] who showed this in the isotropic case. We remark that unconditionality of multiwavelets with arbitrary dilations in $L^p$ space ($1 < p < \infty$) was shown by Pompe [Po].

In the following section we study the sequence space of coefficients of elements of $H^p$ in the wavelet expansion. In the isotropic case this sequence space was studied by Meyer [Me] and Frazier and Jawerth [FJ1, FJ2] in the scale of Triebel-Lizorkin spaces.

**Wavelet preliminaries.** Throughout Section 2, we are going to assume that we have a lattice $\Gamma$ ($\Gamma = P\mathbb{Z}^n$ for some nondegenerate $n \times n$ matrix $P$) and a *dilation* matrix $A$ preserving $\Gamma$, i.e., all eigenvalues $\lambda$ of $A$ satisfy $|\lambda| > 1$, and $A\Gamma \subset \Gamma$. Without loss of generality, we will assume that $\Gamma = \mathbb{Z}^n$.

DEFINITION 1.1. Let $\Psi$ be a finite family of functions $\Psi = \{\psi^1, \ldots, \psi^L\} \subset L^2(\mathbb{R}^n)$. We say that $\Psi$ is a *wavelet family* (or a *multiwavelet*) if $\{\psi_{j,k}^l : j \in \mathbb{Z}, k \in \mathbb{Z}^n, l = 1, \ldots, L\}$ is an orthonormal basis for $L^2(\mathbb{R}^n)$. Here, for $\psi \in L^2(\mathbb{R}^n)$ we use the convention

$$\psi_{j,k}(x) = D_{A^j}\tau_k\psi(x) = |\det A|^{j/2}\psi(A^j x - k) \qquad j \in \mathbb{Z}, k \in \mathbb{Z}^n,$$

where $\tau_y f(x) = f(x-y)$ is a translation operator by the vector $y \in \mathbb{R}^n$, and $D_A f(x) = \sqrt{|\det A|} f(Ax)$ is a dilation operator by the matrix A.

DEFINITION 1.2. By a *multiresolution analysis* we mean a sequence of closed subspaces $(V_i)_{i \in \mathbb{Z}} \subset L^2(\mathbb{R}^n)$ satisfying:
(i) $V_i \subset V_{i+1}$ for $i \in \mathbb{Z}$,
(ii) $V_i = D_{A^i} V_0$ for $i \in \mathbb{Z}$,
(iii) $\overline{\bigcup_{i \in \mathbb{Z}} V_i} = L^2(\mathbb{R}^n)$,
(iv) $\bigcap_{i \in \mathbb{Z}} V_i = \{0\}$,
(v) there exists $\varphi$ called a *scaling function* such that $\{\tau_k \varphi\}_{k \in \mathbb{Z}^n}$ is an orthonormal basis of $V_0$.

We say that a wavelet family $\Psi = \{\psi^1, \ldots, \psi^L\}$ is associated with a multiresolution analysis (MRA), if the spaces

$$(1.1) \qquad V_i = \bigoplus_{j<i} W_j, \qquad \text{where } W_j = \overline{\text{span}}\{\psi_{j,k}^l : k \in \mathbb{Z}^n, l = 1, \ldots, L\},$$

form an MRA. This happens precisely when $V_0$ as a shift invariant subspace of $L^2(\mathbb{R}^n)$ has dimension function $D_\Psi(\xi) = 1$ for a.e. $\xi \in \mathbb{R}^n$. For the definition of the dimension function for general shift invariant spaces, see [BDR, Bo3]. However, the dimension function of $V_0$ is given by the explicit formula,

$$D_\Psi(\xi) = \sum_{l=1}^{L} \sum_{j=1}^{\infty} \sum_{k \in \mathbb{Z}^n} |\hat{\psi}^l(B^j(\xi+k))|^2, \tag{1.2}$$

where $B = A^T$, see [Ba, BRS]. Since

$$\int_{(0,1)^n} D_\Psi(\xi) d\xi = 1/(b-1) \sum_{l=1}^{L} ||\psi_l||^2, \tag{1.3}$$

a wavelet family $\Psi = \{\psi^1, \ldots, \psi^L\}$ can be associated with an MRA only if $L = b-1$, where $b = |\det A|$.

DEFINITION 1.3. We say that a function $f$ on $\mathbb{R}^n$ is $r$-regular, if $f$ is of class $C^r$, $r = 0, 1, \ldots, \infty$ and

$$|\partial^\alpha f(x)| \leq c_{\alpha,k}(1+|x|)^{-k}, \tag{1.4}$$

for each $k \in \mathbb{N}$, and each multi-index $\alpha$, with $|\alpha| \leq r$. A wavelet family $\Psi = \{\psi^1, \ldots, \psi^L\}$ is $r$-regular, if $\psi^1, \ldots, \psi^L$ are $r$-regular functions. An $MRA$ is $r$-regular if the subspace $V_0$ given by (1.1) has an orthonormal basis of the form $\{\tau_k \varphi : k \in \mathbb{Z}^n\}$ for some $r$-regular scaling function $\varphi$.

If a wavelet family $\Psi$ is $r$-regular for $r$ sufficiently large, or more precisely $|\hat{\psi}^l(\xi)|$ are continuous and $|\hat{\psi}^l(\xi)| \leq C(1+|\xi|)^{-n/2-\varepsilon}$ for some $\varepsilon > 0$, then the sum (1.2) converges uniformly on compact subsets of $\mathbb{R}^n \setminus \mathbb{Z}^n$ to the continuous function $D_\Psi(\xi)$ having integer values. Therefore, by $\mathbb{Z}^n$-periodicity $D_\Psi$ is constantly equal $d$ for some $d \in \mathbb{N}$. If $d = 1$ then $r$-regular wavelet family $\Psi$ comes from some MRA (more generally, $\Psi$ comes from an MRA with multiplicity $d$). This result was essentially shown by Auscher [Au2, Au3, Theorem 10.1] (under slightly stronger assumptions on $\hat{\psi}^l$'s). In general, we can not expect that this MRA is also $r$-regular; for a counterexample see [MC, Chapter 8, Proposition 2]. Conversely, having an $r$-regular MRA we can not, in general, deduce the existence of $r$-regular wavelet family associated with it, see [Wo2, Theorem 5.10, Remark 5.6]. Nevertheless, we can deduce the existence of $r$-regular wavelet family by using the following result, see [Wo2, Corollary 5.17] which also holds for $r = \infty$.

PROPOSITION 1.4 (Wojtaszczyk). *Assume that we have a multiresolution analysis on $\mathbb{R}^n$ associated with an integral dilation $A$ with $|\det A| = b$. Assume that this MRA has an $r$-regular scaling function $\varphi(x)$ such that $\hat{\varphi}(\xi)$ is real for some $r = 0, 1, \ldots, \infty$. Then there exists a wavelet family associated with this MRA consisting of a $(b-1)$ $r$-regular function.*

Starting in Section 3 we relax the assumption that a dilation $A$ has integer entries and we only assume that $A$ is expansive in the sense of Definition 2.1 in Chapter 1. In Section 3 we show that in this setting $r$-regular orthogonal wavelet bases may not exist in general. Consequently, starting in Section 4 we consider also tight frame wavelets.

## 2. Wavelets in the Schwartz class

In this section we are going to construct an MRA which has a scaling function $\varphi$ in the Schwartz class and $\hat{\varphi}(\xi)$ is real for some special class of dilations satisfying a kind of expansiveness property. By Proposition 1.4, we can then find a wavelet family in the Schwartz class associated with this MRA.

DEFINITION 2.1. We say that the integral dilation $B$ is *strictly expansive* if there exists a compact set $K \subset \mathbb{R}^n$ such that
- $0 \in K^\circ$, where $K^\circ$ is the interior of $K$,
- $|K \cap (l + K)| = \delta_{l,0}$ for $l \in \mathbb{Z}^n$,
- $K \subset BK^\circ$.

DEFINITION 2.2. Given a set $Y \subset \mathbb{R}^n$ and $\varepsilon > 0$ we define its $\varepsilon$-*interior* $Y^{-\varepsilon}$ and $\varepsilon$-*neighborhood* $Y^{+\varepsilon}$ by

$$Y^{-\varepsilon} = \{\xi \in \mathbb{R}^n : \mathbf{B}(\xi,\varepsilon) \subset Y\},$$
$$Y^{+\varepsilon} = \{\xi \in \mathbb{R}^n : \mathbf{B}(\xi,\varepsilon) \cap Y \neq \emptyset\}.$$

Note that $Y^{-\varepsilon}$ is closed, $Y^{+\varepsilon}$ is open, and the interior of $Y$ satisfies $Y^\circ = \bigcup_{\varepsilon>0} Y^{-\varepsilon}$. If the dilation $B$ is strictly expansive and the compact set $K$ satisfies Definition 2.1, then there exists $\varepsilon > 0$ such that

(2.1) $$K^{+\varepsilon} \subset B(K^{-\varepsilon}).$$

We shall prove the following existence theorem.

THEOREM 2.3. *Suppose $A$ is a dilation matrix with $A\mathbb{Z}^n \subset \mathbb{Z}^n$ with $b = |\det A|$. If the dilation $B = A^T$ is strictly expansive then there exists a multiresolution analysis with a scaling function and an associated wavelet family of $(b-1)$ functions in the Schwartz class.*

For the sake of completeness, we recall the proof of Theorem 2.3 from [BS1, Theorem 3.2].

PROOF. Let $K$ and $\varepsilon > 0$ satisfy Definition 2.1 and (2.1). Choose a $C^\infty$ function $g : \mathbb{R}^n \to [0,\infty)$ such that $\int_{\mathbb{R}^n} g = 1$ and

(2.2) $$\operatorname{supp} g := \{\xi \in \mathbb{R}^n : g(\xi) \neq 0\} = \mathbf{B}(0,\varepsilon).$$

Define the function $f$ by

(2.3) $$f(\xi) = (\mathbf{1}_K * g)(\xi).$$

Clearly $f$ is in the class $C^\infty$, $0 \leq f(\xi) \leq 1$, and

(2.4) $$\operatorname{supp} f = \{\xi \in \mathbb{R}^n : f(\xi) \neq 0\} \subset K^{+\varepsilon},$$
(2.5) $$\{\xi \in \mathbb{R}^n : f(\xi) = 1\} = K^{-\varepsilon}.$$

Moreover,

(2.6) $$\sum_{k \in \mathbb{Z}^n} f(\xi+k) = \sum_{k \in \mathbb{Z}^n} \int_{\mathbb{R}^n} \mathbf{1}_K(\xi+k-\eta)g(\eta)d\eta = 1 \quad \text{for all } \xi \in \mathbb{R}^n,$$

since $\sum_{k \in \mathbb{Z}^n} \mathbf{1}_K(\xi+k) = 1$ for a.e. $\xi \in \mathbb{R}^n$ by Definition 2.1.

Finally, define the function $m : \mathbb{R}^n \to [0,1]$ by

$$(2.7) \qquad m(\xi) = \sqrt{\sum_{k \in \mathbb{Z}^n} f(B(\xi + k))}.$$

CLAIM 2.4. *The function $m$ given by (2.7) is $C^\infty$, $\mathbb{Z}^n$-periodic, and*

$$(2.8) \qquad \sum_{d \in \mathcal{D}} |m(\xi + B^{-1}d)|^2 = 1 \qquad \text{for all } \xi \in \mathbb{R}^n,$$

$$(2.9) \qquad m(\xi) > 0 \implies \xi \in \mathbb{Z}^n + B^{-1}(K^{+\varepsilon}),$$

$$(2.10) \qquad m(\xi) = 0 \qquad \text{for } \xi \in (B^{-1}\mathbb{Z}^n \setminus \mathbb{Z}^n) + B^{-1}(K^{-\varepsilon}),$$

*where $B = A^T$, and $\mathcal{D} = \{d_1, \ldots, d_b\}$ is the set of representatives of different cosets of $\mathbb{Z}^n / B\mathbb{Z}^n$, where $b = |\det A|$.*

PROOF OF CLAIM 2.4. To guarantee that $m$ is $C^\infty$, the function $f$ must "vanishes strongly", i.e., if $f(\xi_0) = 0$ for some $\xi_0$ then $\delta^\alpha f(\xi_0) = 0$ for any multi-index $\alpha$. It is clear that if nonnegative function $f$ in $C^\infty$ "vanishes strongly" then $\sqrt{f}$ is also $C^\infty$.

The condition (2.8) is a consequence of

$$\sum_{d \in \mathcal{D}} |m(\xi + B^{-1}d)|^2 = \sum_{k \in \mathbb{Z}^n} \sum_{d \in \mathcal{D}} f(B(\xi + B^{-1}d + k)) = \sum_{k \in \mathbb{Z}^n} \sum_{d \in \mathcal{D}} f(\xi + d + Bk) = 1,$$

by (2.6).

To see (2.9), take $\xi$ such that $m(\xi) > 0$. By (2.4) and (2.7), $B(\xi + k) \in K^{+\varepsilon}$ for some $k \in \mathbb{Z}^n$, and hence (2.9) holds.

We claim that (2.10) follows from (2.8) and

$$(2.11) \qquad m(\xi) = 1 \qquad \text{for } \xi \in \mathbb{Z}^n + B^{-1}(K^{-\varepsilon}).$$

Indeed, if $\xi \in B^{-1}d + k + B^{-1}(K^{-\varepsilon})$ for some $d \in \mathcal{D} \setminus B\mathbb{Z}^n$ and $k \in \mathbb{Z}^n$, then by (2.11) we have $m(\xi - B^{-1}d) = 1$. Hence by (2.8) $m(\xi) = 0$ and (2.10) holds. Finally, (2.11) is the immediate consequence of (iii) and (2.7). This ends the proof of the claim. $\square$

We can write $m$ in the Fourier expansion as

$$(2.12) \qquad m(\xi) = \frac{1}{\sqrt{|\det A|}} \sum_{k \in \mathbb{Z}^n} h_k e^{-2\pi i \langle k, \xi \rangle},$$

where we include the factor $|\det A|^{-1/2}$ outside the summation as in [Bo1]. Since $m$ is $C^\infty$, the coefficients $h_k$ decay polynomially at infinity, that is for all $N > 0$ there is $C_N > 0$ so that

$$|h_k| \leq C_N |k|^{-N} \qquad \text{for } k \in \mathbb{Z}^n \setminus \{0\}.$$

Since $m$ satisfies (2.8) and $m(0) = 1$, $m$ is a low-pass filter which is regular in the sense of the definition following [Bo1, Theorem 1]. By [Bo1, Theorem 5] $\varphi \in L^2(\mathbb{R}^n)$ defined by

$$(2.13) \qquad \hat{\varphi}(\xi) = \prod_{j=1}^{\infty} m(B^{-j}\xi),$$

has orthogonal translates, i.e.,
$$\langle \varphi, \tau_l \varphi \rangle = \delta_{l,0} \quad \text{for } l \in \mathbb{Z}^n,$$
if and only if $m$ satisfies the Cohen condition, that is there exists compact set $\tilde{K} \subset \mathbb{R}^n$ such that
- $\tilde{K}$ contains a neighborhood of zero,
- $|\tilde{K} \cap (l + \tilde{K})| = \delta_{l,0}$ for $l \in \mathbb{Z}^n$,
- $m(B^{-j}\xi) \neq 0$ for $\xi \in \tilde{K}$, $j \geq 1$.

The first guess for $\tilde{K}$ to be $K$ is in general incorrect, e.g. if $K$ has isolated points. Instead we claim that there is $0 < \delta < 1$ so that

(2.14) $$\tilde{K} = \{\xi \in K : |\mathbf{B}(\xi, \varepsilon) \cap K| \geq \delta |\mathbf{B}(\xi, \varepsilon)|\}$$

does the job. Clearly, if $\xi \in \tilde{K}$ then $h(\xi) \neq 0$, hence $f(\xi) \neq 0$ and thus $m(B^{-1}\xi) \neq 0$. By (2.1) $B^{-1}\tilde{K} \subset B^{-1}K^{+\varepsilon} \subset K^{-\varepsilon} \subset \tilde{K}$ and thus $m(B^{-j}\xi) \neq 0$ for all $j \geq 1$. Finally, it suffices to check that

(2.15) $$\sum_{k \in \mathbb{Z}^n} \mathbf{1}_{\tilde{K}}(\xi + k) \geq 1 \quad \text{for all } \xi \in \mathbb{R}^n.$$

By the compactness of $K$ there is a finite index set $I \subset \mathbb{Z}^n$ such that

(2.16) $$\sum_{k \in I} \mathbf{1}_K(\xi + k) \geq 1 \quad \text{for all } \xi \in [-1,1]^n.$$

Take any $\xi \in [-1/2, 1/2]^n$ and integrate (2.16) over $\mathbf{B}(\xi, \varepsilon)$ to obtain
$$\sum_{k \in I} |\mathbf{B}(\xi + k, \varepsilon) \cap K| \geq |\mathbf{B}(\xi, \varepsilon)|.$$

Therefore, if we take $\delta = 1/\#I$ then there is $k \in I$ such that $|\mathbf{B}(\xi + k, \varepsilon) \cap K| \geq \delta |\mathbf{B}(\xi, \varepsilon)|$ and hence $\xi + k \in \tilde{K}$. Thus, (2.15) holds and $\tilde{K}$ given by (2.14) satisfies the Cohen condition. Therefore, $\varphi$ is a scaling function for the multiresolution analysis $(V_j)_{j \in \mathbb{Z}}$ defined by
$$V_j = \overline{\text{span}}\{D_{A^j}\tau_l \varphi : l \in \mathbb{Z}^n\} \quad \text{for } j \in \mathbb{Z}.$$

It remains to show that $\varphi \in \mathcal{S}$. We are going to prove that $\varphi$ is band-limited, i.e., $\hat{\varphi}$ is compactly supported. By (2.9) and (2.13)

(2.17) $$\hat{\varphi}(\xi) \neq 0 \implies \xi \in B\mathbb{Z}^n + K^{+\varepsilon}.$$

On the other hand, by (2.10) $m(B^{-j}\xi) = 0$ for $\xi \in B^{j-1}\mathbb{Z}^n \setminus B^j\mathbb{Z}^n + B^{j-1}(K^{-\varepsilon})$. Since
$$\bigcup_{j=2}^{\infty} (B^{j-1}\mathbb{Z}^n \setminus B^j\mathbb{Z}^n) = B\mathbb{Z}^n \setminus \{0\},$$
and
$$K^{+\varepsilon} \subset B(K^{-\varepsilon}) \subset B^{j-1}(K^{-\varepsilon}) \quad \text{for } j \geq 2,$$
we have

(2.18) $$\hat{\varphi}(\xi) = 0 \quad \text{for } \xi \in B\mathbb{Z}^n \setminus \{0\} + K^{+\varepsilon}.$$

Combining (2.17) and (2.18) we have $\hat{\varphi}(\xi) = 0$ for $\xi \in (K^{+\varepsilon})^{\mathbf{c}}$. Therefore, $\operatorname{supp} \hat{\varphi} \subset K^{+\varepsilon}$ and $\varphi$ is in the Schwartz class. To conclude the proof it suffices to use Proposition 1.4 for $r = \infty$. $\square$

As a corollary of Theorem 2.3 we have the following.

COROLLARY 2.5. *Suppose $A$ is a dilation with integer entries with $b = |\det A|$. Then there exists $m \in \mathbb{N}$ and a multiresolution analysis with a scaling function and a wavelet family of $(b^m - 1)$ functions in the Schwartz class associated to the dilation $A^m$.*

PROOF. It suffices to notice that $B^m = (A^T)^m$ is strictly expansive for sufficiently large $m \in \mathbb{N}$ and $K = [-1/2, 1/2]^n$. $\square$

Even though the number of functions in the multiwavelet in the above corollary could be much bigger than $b-1$, Corollary 2.5 still has great significance. By results of Section 5, a multiwavelet $\Psi$ in the Schwartz class generates an unconditional basis of $H_A^p = H_{A^m}^p$ for the whole range of $0 < p < \infty$. This can not be achieved if we use an $r$-regular multiwavelet instead. Finally, note that Corollary 2.5 also holds for a larger class of dilations, e.g., dilations $A$ such that some positive power of $A$ has integer entries.

It is not known whether there exist orthonormal wavelets in the Schwartz class for dilations with integer entries that do not necessarily satisfy the strict expansiveness property. Speegle and the author [BS1] showed that one can construct an $\infty$-regular MRA associated with $\infty$-regular wavelet family for all $2 \times 2$ dilations with integer entries. Since the case $n = 1$ is trivial, this problem remains open in dimensions $n \geq 3$.

Nevertheless, one can always show that there exist $r$-regular multiwavelets for $r < \infty$ and general dilations with integer entries.

THEOREM 2.6. *Suppose $A$ is a dilation with integer entries. For every $r \in \mathbb{N}$ there exists an $r$-regular multiresolution analysis and an associated $r$-regular wavelet family of $(|\det A| - 1)$ functions.*

Theorem 2.6 was shown by Strichartz [Sr] under an additional hypothesis that $A$ admits a self-affine tiling, see [GM, LW1–LW3]. This assumption was removed by the author; the proof of Theorem 2.6 can be found in [Bo4].

Finally, we mention that Daubechies' construction of arbitrarly smooth compactly supported wavelets was extended to certain non-dyadic dilations in higher dimensions, see [Ay1, Ay2, BW].

## 3. Limitations on orthogonal wavelets

The construction of Section 2 applies only when the dilation $A$ has integer entries, or more generally, $A$ preserves some lattice $\Gamma = P\mathbb{Z}^n$, where $P$ is an $n \times n$ nonsingular matrix. If the dilation $A$ does not preserve any lattice then there could be no well localized wavelets, even 0-regular. Chui and Shi [CS] have shown that all orthogonal wavelets associated with "almost any" irrational dilation in the dimension $n = 1$ must be MSF (minimally supported in frequency). We are going to show that the analogous statement holds for general multiwavelets.

We say that a multiwavelet $\Psi = \{\psi^1, \ldots, \psi^L\}$ associated with $A$ is MSF, if $|\hat\psi^l| = \mathbf{1}_{W_l}$ for some measurable sets $W_l$. By [BRS, Theorem 2.4] these sets are characterized by

(3.1)
$$\sum_{k \in \mathbb{Z}^n} \mathbf{1}_{W_l}(\xi+k)\mathbf{1}_{W_{l'}}(\xi+k) = \delta_{l,l'} \qquad \text{a.e. } \xi \in \mathbb{R}^n, \quad l,l' = 1,\ldots,L,$$
$$\sum_{j \in \mathbb{Z}} \sum_{l=1}^{L} \mathbf{1}_{W_l}(B^j \xi) = 1 \qquad \text{a.e. } \xi \in \mathbb{R}^n,$$

where $B = A^T$. We say that a multiwavelet $\Psi = \{\psi^1, \ldots, \psi^L\}$ associated with $A$ is *combined MSF* if

(3.2) $$\sum_{l=1}^{L} |\hat\psi^l(\xi)|^2 = \mathbf{1}_W(\xi) \qquad \text{for a.e. } \xi \in \mathbb{R}^n,$$

for some multiwavelet set $W$ of order $L$, i.e., $W = \bigcup_{l=1}^{L} W_l$ for some $W_1, \ldots, W_L$ satisfying (3.1). By [BRS, Theorem 2.6] a multiwavelet set $W$ of order $L$ is characterized by

(3.3) $$\sum_{k \in \mathbb{Z}^n} \mathbf{1}_W(\xi + k) = L \qquad \text{a.e. } \xi \in \mathbb{R}^n,$$

(3.4) $$\sum_{j \in \mathbb{Z}} \mathbf{1}_W(B^j \xi) = 1 \qquad \text{a.e. } \xi \in \mathbb{R}^n,$$

where $B = A^T$.

THEOREM 3.1. *Suppose that a dilation $A$ is such that for every integer $j \geq 1$ the rows of $A^{-j}$ (treated as vectors in $\mathbb{R}^n$) together with the standard basis vectors $e_1, \ldots, e_n$ are linearly independent over $\mathbb{Q}$. Then any orthogonal multiwavelet $\Psi = \{\psi^1, \ldots, \psi^L\}$ associated with $A$ is combined MSF, i.e., (3.2) holds for some set $W$ satisfying (3.3) and (3.4).*

PROOF. By the orthogonality

$$\delta_{l,l'}\delta_{j,0}\delta_{k,k'} = \langle \psi^l_{j,k}, \psi^{l'}_{0,k'}\rangle = b^{j/2}\int_{\mathbb{R}^n} \psi^l(A^j x - k)\overline{\psi^{l'}(x-k')}dx$$
$$= b^{j/2}\int_{\mathbb{R}^n}\psi^l(A^j x + A^j k' - k)\overline{\psi^{l'}(x)}dx$$

for all $j \in \mathbb{Z}$, $k, k' \in \mathbb{Z}^n$, $l, l' = 1, \ldots, L$. By Plancherel's formula

$$
\begin{aligned}
(3.5) \quad 0 &= \int_{\mathbb{R}^n} \overline{\hat{\psi}^{l'}(\xi)} \hat{\psi}^l(B^{-j}\xi) e^{2\pi i \langle B^{-j}\xi, A^j k' - k \rangle} d\xi \\
&= \int_{\mathbb{R}^n} \overline{\hat{\psi}^{l'}(\xi)} \hat{\psi}^l(B^{-j}\xi) e^{2\pi i \langle \xi, k' - A^{-j}k \rangle} d\xi.
\end{aligned}
$$

By Lemma 3.2 we conclude that $\mathbb{Z}^n + A^{-j}\mathbb{Z}^n$ is dense in $\mathbb{R}^n$ for all $j \geq 1$. Therefore, by the Fourier Inversion Formula and (3.5), $\hat{\psi}^{l'}(\xi)\hat{\psi}^l(B^{-j}\xi) = 0$ for a.e. $\xi \in \mathbb{R}^n$ and all $j \geq 1$, $l, l' = 1, \ldots, L$. Let $W = \bigcup_{l=1}^{L} \operatorname{supp} \hat{\psi}^l$. We have $|W \cap B^j W| = 0$ for $j \geq 1$, and hence for all $j \in \mathbb{Z} \setminus \{0\}$. Now (3.4) holds because the multiwavelet $\Psi$ satisfies the discrete Calderón formula $\sum_{l=1}^{l} |\hat{\psi}^l(B^j \xi)|^2 = 1$ for a.e. $\xi$, see [Bo2, CMW]. For $j \in \mathbb{Z}$ let

$$W_j = \overline{\operatorname{span}}\{\psi_{j,k}^l : k \in \mathbb{Z}^n, \ l = 1, \ldots, L\}.$$

Naturally we have $W_0 \subset \{f \in L^2 : \operatorname{supp} \hat{f} \subset W\}$. But by (3.4) and $\bigoplus_{j \in \mathbb{Z}} W_j = L^2(\mathbb{R}^n)$ we must have

$$(3.6) \qquad W_0 = \{f \in L^2 : \operatorname{supp} \hat{f} \subset W\}.$$

The range function of a shift invariant space $W_0$ is given by

$$J(\xi) = \operatorname{span}\{(\hat{\psi}^l(\xi + k))_{k \in \mathbb{Z}^n} : l = 1, \ldots, L\} \subset \ell^2(\mathbb{Z}^n),$$

see [BDR, Bo3, Proposition 1.5]. Also by [Bo3, Proposition 1.5] and (3.6),

$$J(\xi) = \{(s_k)_{k \in \mathbb{Z}^n} \in \ell^2(\mathbb{Z}^n) : s_k = 0 \text{ for } \xi + k \notin W\}.$$

Since the vectors $(\hat{\psi}^l(\xi + k))_{k \in \mathbb{Z}^n}$, $l = 1, \ldots, L$ form an orthonormal basis of $J(\xi)$ we have

$$\sum_{l=1}^{L} \langle s, (\hat{\psi}^l(\xi + m))_{m \in \mathbb{Z}^n} \rangle (\psi^l(\xi + k))_{k \in \mathbb{Z}^n} = s \qquad \text{for all } s \in J(\xi).$$

By taking $s = \boldsymbol{e}_k \in J(\xi)$, where $\xi + k \in W$ and $\boldsymbol{e}_k$'s are the standard basis vectors of $\ell^2(\mathbb{Z}^n)$, we conclude that $\sum_{l=1}^{L} |\hat{\psi}^l(\xi + k)|^2 = 1$ for $\xi + k \in W$. This shows (3.2). (3.3) is a consequence of (3.2) since $\sum_{k \in \mathbb{Z}^n} |\hat{\psi}^l(\xi + k)|^2 = 1$ for a.e. $\xi$. $\square$

It is worth noting that the above argument may be simplified if one applies the spectral function introduced by Rzeszotnik [BR, Rz].

LEMMA 3.2. *Suppose $D$ is $n \times n$ real matrix such that the rows of $D$ (treated as vectors in $\mathbb{R}^n$) together with vectors $e_1, \ldots, e_n$ are linearly independent over $\mathbb{Q}$. Then the set $\mathbb{Z}^n + D\mathbb{Z}^n$ is dense in $\mathbb{R}^n$.*

Since the author does not know an elementary proof of this result we are going to show something more, namely that that a sequence obtained by a certain ordering of $D\mathbb{Z}^n$ is uniformly distributed mod 1. We are going to follow the excellent book of Kuipers and Niederreiter [KN].

DEFINITION 3.3. Suppose $\boldsymbol{a} = (a_1, \ldots, a_n)$, $\boldsymbol{b} = (b_1, \ldots, b_n) \in \mathbb{R}^n$. We say that $\boldsymbol{a} < \boldsymbol{b}$ ($\boldsymbol{a} \leq \boldsymbol{b}$) if $a_j < b_j$ ($a_j \leq b_j$) for all $j = 1, \ldots, n$. The cube $[\boldsymbol{a}, \boldsymbol{b})$ is defined as $\{\boldsymbol{x} \in \mathbb{R}^n : \boldsymbol{a} \leq \boldsymbol{x} < \boldsymbol{b}\}$. The *fractional part* of $\boldsymbol{x} \in \mathbb{R}^n$ is $\{\boldsymbol{x}\} = (\{x_1\}, \ldots, \{x_n\})$, where $\{x\} = x - \lfloor x \rfloor$. Given a sequence $(\boldsymbol{x}_k)_{k \in \mathbb{N}} \subset \mathbb{R}^n$, $E \subset I^n = [0,1)^n$, and $N \in \mathbb{N}$, let $\#(E; N)$ denote the number of points $\boldsymbol{x}_k \in E$, $1 \leq k \leq N$. We say that a sequence $(\boldsymbol{x}_k)_{k \in \mathbb{N}} \subset \mathbb{R}^n$ is *uniformly distributed mod 1* if

$$(3.7) \qquad \lim_{N \to \infty} \frac{\#([\boldsymbol{a}, \boldsymbol{b}); N)}{N} = \prod_{j=1}^{n} (b_j - a_j)$$

for all intervals $[\boldsymbol{a}, \boldsymbol{b}) \subset I^n = [0,1)^n$.

The Weyl Criterion, see [KN, Theorem 6.2, Chapter 1], says that a sequence $(\boldsymbol{x}_k)_{k \in \mathbb{N}} \subset \mathbb{R}^n$ is u. d. mod 1 if and only if for every $\boldsymbol{h} \in \mathbb{Z}^n \setminus \{\boldsymbol{0}\}$,

$$(3.8) \qquad \lim_{N \to \infty} \frac{1}{N} \sum_{k=1}^{N} e^{2\pi i \langle \boldsymbol{h}, \boldsymbol{x}_k \rangle} = 0.$$

Therefore, a sequence $(\boldsymbol{x}_k)_{k \in \mathbb{N}} \subset \mathbb{R}^n$ is u. d. mod 1 if and only if for every $\boldsymbol{h} \in \mathbb{Z}^n \setminus \{\boldsymbol{0}\}$, the sequence of real numbers $(\langle \boldsymbol{h}, \boldsymbol{x}_k \rangle)_{k \in \mathbb{N}}$ is u. d. mod 1. By the Weyl Criterion, see [KN, Example 6.1, Chapter 1], given $\boldsymbol{\theta} = (\theta_1, \ldots, \theta_m)$ with the property that the real numbers $1, \theta_1, \ldots, \theta_m$ are linearly independent over $\mathbb{Q}$, the sequence $(k\boldsymbol{\theta})_{k \in \mathbb{N}}$ is u. d. mod 1.

We are now ready to present the proof of Lemma 3.2.

PROOF OF LEMMA 3.2. Given $\boldsymbol{k} = (k_1, \ldots, k_n) \in \mathbb{R}^n$ let $||\boldsymbol{k}||_\infty = \max(|k_1|, \ldots, |k_n|)$. Let $\boldsymbol{k}_1, \boldsymbol{k}_2, \ldots$ be the ordering of all elements of $\mathbb{Z}^n$ such that $||\boldsymbol{k}_i||_\infty < ||\boldsymbol{k}_j||_\infty$ implies $i < j$. It suffices to show that the sequence $(\boldsymbol{x}_i)_{i \in \mathbb{N}} = (D\boldsymbol{k}_i)_{i \in \mathbb{N}}$ is u. d. mod 1. By (3.7) it suffices to show that

$$\lim_{N \to \infty} \frac{\#([\boldsymbol{a}, \boldsymbol{b}); (2N+1)^n)}{(2N+1)^n} = \prod_{j=1}^{n} (b_j - a_j)$$

for all intervals $[\boldsymbol{a}, \boldsymbol{b}) \subset I^n = [0,1)^n$. Therefore, by the Weyl Criterion (3.8) we must show that for every $\boldsymbol{h} \in \mathbb{Z}^n \setminus \{\boldsymbol{0}\}$,

$$(3.9) \qquad \lim_{N \to \infty} \frac{1}{(2N+1)^n} \sum_{\substack{\boldsymbol{k} \in \mathbb{Z}^n \\ ||\boldsymbol{k}||_\infty \leq N}} e^{2\pi i \langle \boldsymbol{h}, D\boldsymbol{k} \rangle} = 0.$$

Fix any $\boldsymbol{h} \in \mathbb{Z}^n \setminus \{\boldsymbol{0}\}$. By our hypothesis there exists $m_0 = 1, \ldots, n$ such that $\theta = \sum_{j=1}^{n} h_j d_{j,m_0} \notin \mathbb{Q}$, where $D = (d_{j,m})_{j=1,\ldots,n}^{m=1,\ldots,n}$. Otherwise we would have $\sum_{j=1}^{n} h_j (d_{j,1}, \ldots, d_{j,n}) \in \mathbb{Q}^n$, which contradicts the linear independence of the rows of $D$ together with the standard basis vectors of $\mathbb{R}^n$ over $\mathbb{Q}$. Since the sequence $(k\theta)_{k \in \mathbb{N}}$ is u. d. mod 1, for every $\varepsilon > 0$ there is $M$ such that for all $N \geq M$

$$(3.10) \qquad \left| \sum_{k=-N}^{N} \frac{1}{2N+1} e^{2\pi i \theta k} \right| < \varepsilon.$$

We also have

$$\frac{1}{(2N+1)^n} \sum_{\substack{\boldsymbol{k} \in \mathbb{Z}^n \\ ||\boldsymbol{k}||_\infty \leq N}} e^{2\pi i \langle \boldsymbol{h}, \boldsymbol{D}\boldsymbol{k} \rangle}$$

$$= \sum_{k_1=-N}^{N} \cdots \sum_{k_n=-N}^{N} \frac{1}{(2N+1)^n} e^{2\pi i \sum_{j=1}^{n} \sum_{m=1}^{n} h_j d_{j,m} k_m}$$

$$= \sum_{k_1,\ldots \check{k}_{m_0} \ldots, k_n=-N}^{N} \frac{1}{(2N+1)^{n-1}} e^{2\pi i \sum_{j=1}^{n} \sum_{m \neq m_0} h_j d_{j,m} k_m} \sum_{k_{m_0}=-N}^{N} \frac{1}{2N+1} e^{2\pi i \theta k_{m_0}}$$

where $\check{k}_{m_0}$ means we omit the index $k_{m_0}$. Therefore, by (3.10) we have

$$\left| \frac{1}{(2N+1)^n} \sum_{\substack{\boldsymbol{k} \in \mathbb{Z}^n \\ ||\boldsymbol{k}||_\infty \leq N}} e^{2\pi i \langle \boldsymbol{h}, \boldsymbol{D}\boldsymbol{k} \rangle} \right| < \varepsilon \qquad \text{for all } N \geq M.$$

Since $\varepsilon > 0$ is arbitrary this shows (3.9) and ends the proof of the lemma. $\square$

REMARKS. It is not hard to see that linear independence over $\mathbb{Q}$ of the rows of $A^{-j}$ together with basis vectors $e_1, \ldots, e_n$ is equivalent to $B^{-j}\mathbb{Q}^n \cap \mathbb{Q}^n = \{0\}$, where $B = A^T$. This, in turn, is equivalent to $B^{-j}\mathbb{Z}^n \cap \mathbb{Z}^n = \{0\}$. Therefore, the hypothesis of Theorem 3.1 can be written in a short form as

(3.11) $$(A^T)^j \mathbb{Z}^n \cap \mathbb{Z}^n = \{0\} \qquad \text{for all } j \in \mathbb{Z} \setminus \{0\}.$$

Combined MSF multiwavelets are not well localized in the direct domain. Indeed, if $\psi^l \in L^1$ for all $l = 1, \ldots, L$ then $\hat{\psi}^l$ is continuous and (3.2) can never hold. Hence, combined MSF multiwavelets are of little use in other spaces than $L^2$. A notable exception is the Shannon wavelet $\hat{\psi}(\xi) = \mathbf{1}_{[-1,-1/2] \cup [1/2,1]}$, which generates an unconditional basis for $L^p$ spaces with $1 < p < \infty$, see [Wo4]. Therefore, Theorem 3.1 says then that orthogonal multiwavelets associated with "most" dilations have some limitations in applications to other function spaces than $L^2$.

It also turns out that biorthogonal wavelets are not a good surrogate. Indeed, a slight modification of the proof of Theorem 3.1 shows that all biorthogonal multiwavelets $\Psi = \{\psi^1, \ldots, \psi^L\}$ associated with a large class of dilations, satisfying the hypothesis of Theorem 3.1, must necessarily be combined MSF. As a consequence,

$$a \mathbf{1}_W(\xi) \leq \sum_{l=1}^{L} |\hat{\psi}^l(\xi)|^2 \leq c \mathbf{1}_W(\xi) \qquad \text{for a.e. } \xi \in \mathbb{R}^n,$$

for some multiwavelet set $W$ of order $L$ and $0 < a \leq c < \infty$. This can be shown by adapting the proof of Theorem 3.1; for an alternative approach, see [Bo5]. Therefore, biorthogonal wavelets associated with dilations satisfying (3.11) can not be well localized in the direct domain.

Moreover, Speegle and the author showed that Theorem 3.1 is sharp in the sense that it has a converse. For the proof of Theorem 3.3, see [BS2, Theorem 5.4].

THEOREM 3.3. *Suppose $A$ is a dilation such that every orthogonal multiwavelet $\Psi$ associated with $A$ is combined MSF. Then $A$ must satisfy the hypothesis of Theorem 3.1, i.e., (3.11) holds.*

Another interesting issue is to determine the class of dilation matrices that allow regular orthogonal wavelets. As we have seen in Theorems 2.6 and 3.1, this class certainly includes all dilations with integer entries and excludes all dilations satisfying (3.11) and thus far from preserving the lattice $\mathbb{Z}^n$. This still leaves a large class of dilations for which it is not known whether there exist regular wavelets.

In one dimension, the above problem goes back to Daubechies [Da2] who asked whether there exist orthonormal wavelet bases with good time-frequency localization for irrational dilation factors $a$. This question was partially answered by Chui and Shi [CS]. The complete answer was given by the author who showed that all orthonormal wavelets associated with irrational dilation factors have poor time-frequency localization, see [Bo6]. Combining this with the construction of $\infty$-regular orthogonal wavelets for rational dilations due to Auscher [Au1, KL], we obtain the complete picture in one dimension.

The higher dimensional version of this problem remains open. It is not even known whether there are regular wavelets for diagonal $2 \times 2$ dilations such that one entry is rational while the other is irrational. Therefore, it is perfectly conceivable that there are dilations $A$ such that all other dilations with the same quasi-norm do not allow regular wavelet bases. A good candidate is the dilation $A = \begin{pmatrix} a & 0 \\ 0 & 2 \end{pmatrix}$, where $a > 0$ satisfies the condition that $\forall r > 0$, $2^r \in \mathbb{Q} \implies a^r \notin \mathbb{Q}$. By Theorem 10.5 in Chapter 1, all dilations $A'$ with $H_A^p = H_{A'}^p$, $0 < p \leq 1$, must be of the form $A' = \begin{pmatrix} \pm a^r & 0 \\ 0 & \pm 2^r \end{pmatrix}$, for some real $r > 0$. Hence, either $a^{rj}$ or $2^{rj}$ is irrational for all integers $j \geq 1$. Thus, it is very likely that there are no regular wavelets associated with such dilations $A'$. This shows the importance of having a substitute for orthonormal multiwavelets.

## 4. Non-orthogonal wavelets in the Schwartz class

The results of Section 3 show the demand for an alternative to orthogonal wavelets. In the theory of wavelets a natural substitute for an orthogonal basis is a frame. In this section we show the existence of tight frame wavelets for all dilations. This fact is probably a part of the folklore.

Frames were originally introduced by Duffin and Schaeffer [DS] to study non-harmonic Fourier series. More recently, frames have found applications in many other areas such as wavelets, Weyl-Heisenberg (Gabor) systems, sampling theory, signal processing, etc. For a good introduction to frames, see [Da2, HeW, Yo]. For a more extensive treatment of frame theory, see [Cz, HL]. We start by recalling the notion of a Bessel family and a frame.

DEFINITION 4.1. A subset $X$ of a Hilbert space $\mathcal{H}$ is a *Bessel family* if there exists $c > 0$ so that

$$(4.1) \qquad \sum_{\eta \in X} |\langle f, \eta \rangle|^2 \leq c \|f\|^2 \qquad \text{for } f \in \mathcal{H}.$$

If in addition there exists $0 < a \leq c$ such that

$$(4.2) \qquad a\|f\|^2 \leq \sum_{\eta \in X} |\langle f, \eta \rangle|^2 \leq c\|f\|^2 \qquad \text{for } f \in L^2(\mathbb{R}^n),$$

then $X$ is a *frame*. A frame is *tight* if $a$, $c$ can be chosen so that $a = c$.

Given a Bessel family $X \subset \mathcal{H}$ we define an *analysis operator* as a mapping $\mathcal{H} \ni f \mapsto (\langle f, \eta \rangle)_{\eta \in X} \in \ell^2(X)$. The dual of this map is the *synthesis operator* mapping $(c_\eta)_{\eta \in X} \in \ell^2(X) \mapsto \sum_{\eta \in X} c_\eta \eta \in \mathcal{H}$, where the series converges unconditionally in $\mathcal{H}$.

DEFINITION 4.2. Let $\Psi$ be a finite family of functions $\Psi = \{\psi^1, \ldots, \psi^L\} \subset L^2(\mathbb{R}^n)$. We say that $\Psi$ is a *tight frame multiwavelet* (or a *Bessel multiwavelet*) if $\{\psi_{j,k}^l : j \in \mathbb{Z}, k \in \mathbb{Z}^n, l = 1, \ldots, L\}$ is a tight frame with constant 1 for $L^2(\mathbb{R}^n)$ (or a Bessel family).

It is relatively easy to construct a tight frame multiwavelet for an arbitrary dilation. Here we present a simple construction of multiwavelet consisting of a single function $\psi$ which is in the Schwartz class and $\hat{\psi}$ is $C^\infty$ with compact support based on the example in [Bo2].

THEOREM 4.2. *Given an arbitrary dilation $A$ there is a tight frame wavelet $\psi$ in the Schwartz class.*

PROOF. For $0 < a < (4\|B\|)^{-1}$ consider $\eta : \mathbb{R}^n \to [0, \infty)$ of class $C^\infty$ such that

$$\operatorname{supp} \eta = \{\xi \in \mathbb{R}^n : a < |\xi| < 2a\|B\|\}.$$

It is not hard to give an explicit example of such function. Since the set $\{j \in \mathbb{Z} : a < \|B^j \xi\| < 2\|B\|a\}$ has at least one element for all $\xi \in \mathbb{R}^n \setminus \{0\}$ we conclude that $\tilde{\eta}(\xi) = \sum_{j \in \mathbb{Z}} \eta(B^j \xi) > 0$ for all $\xi \neq 0$ and $\tilde{\eta}$ is $C^\infty$ on $\mathbb{R}^n \setminus \{0\}$. Define $\psi \in L^2(\mathbb{R}^n)$ by $\hat{\psi}(\xi) = \sqrt{\eta(\xi)/\tilde{\eta}(\xi)}$. Clearly,

$$(4.3) \qquad \sum_{j \in \mathbb{Z}} |\hat{\psi}(B^j \xi)|^2 = \sum_{j \in \mathbb{Z}} \eta(B^j \xi)/\tilde{\eta}(B^j \xi) = \sum_{j \in \mathbb{Z}} \eta(B^j \xi)/\tilde{\eta}(\xi) = 1.$$

By [Bo2, Lemma 3.1] we have

$$\sum_{j\in\mathbb{Z}}\sum_{k\in\mathbb{Z}^n}|\langle f,\psi_{j,k}\rangle|^2 = \sum_{j\in\mathbb{Z}}b^j\int_{\mathbb{R}^n}|\hat{f}(B^j\xi)|^2|\hat{\psi}(\xi)|^2 d\xi$$
$$+ \sum_{j\in\mathbb{Z}}b^j\int_{\mathbb{R}^n}\overline{\hat{f}(B^j\xi)}\hat{\psi}(\xi)\Big[\sum_{m\in\mathbb{Z}^n\setminus\{0\}}\hat{f}(B^j(\xi+m))\overline{\hat{\psi}(\xi+m)}\Big]d\xi,$$

(4.4)

for any $f\in\mathcal{D}$, where

$$\mathcal{D} = \{f\in L^2(\mathbb{R}^n) : \hat{f}\in L^\infty(\mathbb{R}^n), \operatorname{supp}\hat{f}\subset K \text{ for some compact } K\subset\mathbb{R}^n\setminus\{0\}\},$$

is a dense subspace of $L^2(\mathbb{R}^n)$. Since $\operatorname{supp}\hat{\psi}\subset \mathbf{B}(0,1/2)\subset(-1/2,1/2)^n$ we have

$$\sum_{j\in\mathbb{Z}}\sum_{k\in\mathbb{Z}^n}|\langle f,\psi_{j,k}\rangle|^2 = \|f\|_2^2 \qquad \text{for all } f\in\mathcal{D},$$

and hence for all $f\in L^2(\mathbb{R}^n)$ by (4.3) and (4.4). Therefore, $\{\psi_{j,k}\}_{j\in\mathbb{Z},k\in\mathbb{Z}^n}$ forms a tight frame with constant 1 in $L^2(\mathbb{R}^n)$. Note that this frame is not an orthogonal basis since $\|\psi\|_2 < 1$. □

## 5. Regular wavelets as an unconditional basis for $H^p$

**Preliminaries in functional analysis.** In this subsection we recall some basic facts about quasi-Banach spaces.

DEFINITION 5.1. Suppose $X$ is a vector space over $\mathbb{K}$ ($\mathbb{K}$ is either $\mathbb{R}$ or $\mathbb{C}$). We say that the map $||\cdot|| : X \to [0, \infty)$ is a *quasi-norm* on $X$ if
(i) $||x|| = 0 \iff x = 0$,
(ii) $||\alpha x|| = |\alpha| \cdot ||x||$ for all $\alpha \in \mathbb{K}$, $x \in X$,
(iii) there is $c > 0$ so that $||x + y|| \leq c \max(||x||, ||y||)$ for all $x, y \in X$.

We say that $(X, ||\cdot||)$ is a *quasi-Banach space* if $||\cdot||$ is a quasi-norm and $X$ equipped with $||\cdot||$ is complete as a metric linear space.

It is well-known, see [KPR], that the (Hausdorff) topological linear space which is *locally bounded*, i.e., it has a bounded neighborhood of the origin, has a quasi-norm. Conversely, the topology associated with any quasi-norm is locally bounded.

Obviously the condition (iii) is equivalent to the more common
(iii) there is $c' > 0$ so that $||x + y|| \leq c'(||x|| + ||y||)$ for all $x, y \in X$.

DEFINITION 5.2. For $0 < p \leq 1$ we say that the quasi-norm $||\cdot||$ on the vector space $X$ is *p-subadditive* if

(5.1) $$||x + y||^p \leq ||x||^p + ||y||^p \quad \text{for } x, y \in X.$$

We say that a $(X, ||\cdot||)$ is *p-convex space* if $||\cdot||$ is $p$-subadditive. If, in addition, $X$ equipped with $||\cdot||$ is complete we say $X$ is a *p-Banach space*. A 1-Banach space is simply called a Banach space.

Naturally, every $p$-subadditive norm on $X$ is a quasi-norm. Surprisingly, this has a converse statement. By the Aoki-Rolewicz Theorem (see [KPR]) every vector space $X$ with a quasi-norm $||\cdot||$ has an equivalent quasi-norm $|||\cdot|||$ which is $p$-subadditive for some $0 < p \leq 1$. Here $p$ satisfies $2^{1/p} = c$, where $c$ is the same as in (iii). Therefore, every quasi-Banach space is $p$-Banach for some $0 < p \leq 1$.

One could define a notion of a basis (sometimes called a Schauder basis) on any $F$-space (a complete metric linear space), see [KPR]. Nevertheless, we are only interested in unconditional bases in quasi-Banach spaces. Here we are following [KLW, Wo3].

DEFINITION 5.3. Let $(e_m, e_m^*)_{m \in M}$ be a *biorthogonal system* in a quasi-Banach space $X$, i.e., we have $e_m \in X$, $e_m^* \in X^*$, and

$$e_m^*(e_s) = \delta_{m,s} \quad \text{for } m, s \in M.$$

The system $(e_m, e_m^*)_{m \in M}$ is an *unconditional basis* in $X$ if for every $x \in X$ the series $\sum_{m \in M} e_m^*(x) e_m$ converges unconditionally to $x$, that is regardless of the ordering of index set $M$. This implies that there exists a constant $K$ such that

(5.2) $$\left\|\sum_{m \in M} \beta_m e_m^*(x) e_m\right\| \leq K \sup_{m \in M} |\beta_m| \cdot ||x|| \quad \text{for all } x \in X.$$

The smallest such constant $K$ is called an *unconditional basis constant* of the system $(e_m, e_m^*)_{m \in M}$.

Since actually the elements $(e_m)_{m \in M}$ determine the functionals $(e_m^*)_{m \in M}$ it is customary to speak about $(e_m)_{m \in M}$ as being an unconditional basis.

The following fact is going to be very useful.

LEMMA 5.4. *For a series $\sum_{m\in\mathbb{N}} x_m$ in a quasi-Banach space $X$, the following are equivalent:*

(i) *the series is unconditionally convergent, i.e., $\sum_{m\in\mathbb{N}} \beta_m x_m$ converges for any choice of $\beta_m \in \{0,1\}$,*

(ii) *for each permutation $\sigma$ of $\mathbb{N}$ the series $\sum_{m\in\mathbb{N}} x_{\sigma(m)}$ is convergent,*

(iii) *for each bounded sequence $(\beta_m)_{m\in\mathbb{N}}$ of scalars the series $\sum_{m\in M} \beta_m x_m$ is convergent. Furthermore, there is a constant $C > 0$ such that*

$$(5.3) \qquad \left\|\sum_{m\in\mathbb{N}} \beta_m x_m\right\| \leq C \sup_{m\in\mathbb{N}} |\beta_m|.$$

PROOF. The equivalence of (i) and (ii) is a well known fact due to Orlicz, see [Ro, Theorem 3.8.2]. (iii) $\implies$ (i) is trivial, whereas (iii) $\implies$ (ii) follows from an argument involving binary expansions.

Indeed, let $\|\cdot\|$ be a $p$-subadditive quasi-norm defining the topology on $X$, $0 < p \leq 1$. For a finite subset $F \subset \mathbb{N}$ denote $x(F) = \sum_{m\in F} x_m$. Without loss of generality we may assume $0 \leq \beta_m \leq 1$. For each $m$ let $\beta_m = \sum_{k=1}^{\infty} \beta_m(k) 2^{-k}$, $\beta_m(k) \in \{0,1\}$ be a dyadic expansion of $\beta_m$. Since the series $\sum_{m\in\mathbb{N}} x_m$ is unconditionally Cauchy, given $\varepsilon > 0$ there exists $N$ such that $\|x(F)\| \leq \varepsilon$ whenever $F \subset \mathbb{N}$ is finite and $\min F \geq N$. For any such $F$ we have

$$\sum_{m\in F} \beta_m x_m = \sum_{k=1}^{\infty} \sum_{m\in F} \beta_m(k) 2^{-k} x_m = \sum_{k=1}^{\infty} 2^{-k} x(F_k),$$

where $F_k = \{m \in F : \beta_m(k) = 1\}$. Therefore,

$$\left\|\sum_{m\in F} \beta_m x_m\right\| \leq \sum_{k=1}^{\infty} 2^{-kp} \|x(F_k)\| \leq \varepsilon \sum_{k=1}^{\infty} 2^{-kp}.$$

Hence, the series $\sum_{m\in\mathbb{N}} \beta_m x_m$ is unconditionally Cauchy. Since the last estimate is independent of the choice $0 \leq \beta_m \leq 1$ we have (5.3). $\square$

**Calderón-Zygmund operators associated with wavelet expansions.** In this subsection we are going to show that a wide class of operators associated with wavelet expansions are Calderón-Zygmund, and hence they are bounded on $H^p$. The first result of this type was shown by Strömberg [Sö] for a particular class of wavelets with exponential decay at infinity. Meyer extended this result to the class of (isotropic) dyadic $r$-regular wavelets, see [Me, Chapter 6]. Our goal is to show that Meyer's approach works also for non-isotropic wavelet expansions for general dilations. The results of this subsection are valid if we replace the orthogonality condition of wavelets by a Bessel condition.

Suppose that $\Psi = \{\psi^1, \ldots, \psi^L\} \subset L^2(\mathbb{R}^n)$ and $\Phi = \{\phi^1, \ldots, \phi^L\}$ are two Bessel multiwavelets, see Section 4. We assume that $\Psi$ and $\Phi$ which are $r$-regular for some $r \in \mathbb{N}$. We will also require that all $\psi^l$'s and $\phi^l$'s have vanishing moments up to a certain, see Theorem 5.6. Define the index set

$$\Lambda = \{(l, j, k) : l = 1, \ldots, L, j \in \mathbb{Z}, k \in \mathbb{Z}^n\}.$$

We consider the operators being a composition of three simple operations of analysis, multiplication of the sequence space, and synthesis. Namely an *analysis operator* is a mapping

$$(5.4) \qquad L^2(\mathbb{R}^n) \ni f \mapsto (\langle f, \psi_{j,k}^l \rangle)_{(l,j,k) \in \Lambda} \in \ell^2(\Lambda).$$

A *multiplication operator* by a sequence of bounded scalars $\epsilon = (\epsilon_{j,k}^l)_{(l,j,k) \in \Lambda} \in \ell^\infty(\Lambda)$ is a mapping

$$(5.5) \qquad \ell^2(\Lambda) \ni (s_{j,k}^l)_{(l,j,k) \in \Lambda} \mapsto (\epsilon_{j,k}^l s_{j,k}^l)_{(l,j,k) \in \Lambda} \in \ell^2(\Lambda).$$

Finally, a *synthesis operator* is a mapping

$$(5.6) \qquad \ell^2(\Lambda) \ni (s_{j,k}^l)_{(l,j,k) \in \Lambda} \mapsto \sum_{(l,j,k) \in \Lambda} s_{j,k}^l \phi_{j,k}^l \in L^2(\mathbb{R}^n).$$

Given two Bessel multiwavelets $\Psi$ and $\Phi$, and a sequence of bounded scalars $\epsilon = (\epsilon_{j,k}^l)_{(j,k) \in \mathbb{Z} \times \mathbb{Z}^n}^{l=1,\ldots,L}$ we define the operator $T_\epsilon : L^2(\mathbb{R}^n) \to L^2(\mathbb{R}^n)$ by

$$(5.7) \qquad T_\epsilon(f) = \sum_{(l,j,k) \in \Lambda} \epsilon_{j,k}^l \langle f, \psi_{j,k}^l \rangle \phi_{j,k}^l.$$

The operator $T_\epsilon$ is bounded as a composition of bounded operators.

LEMMA 5.5. *Suppose $\Psi$ and $\Phi$ are r-regular Bessel multiwavelets. Then for any sequence of scalars $\epsilon = (\epsilon_{j,k}^l)_{(l,j,k) \in \Lambda}$ with $|\epsilon_{j,k}^l| \leq 1$ the operators $T_\epsilon$ given by (5.7) are Calderòn-Zygmund of order $r$ with uniformly bounded constants independent of $\epsilon$. Moreover, the kernels $K_\epsilon(x,y)$ of $T_\epsilon$ satisfy the symmetric (CZ-r) condition, i.e., there exists a constant $C$ such that for every $x \neq y$*

$$(5.8) \qquad |\partial_y^\alpha \partial_x^\beta [K_\epsilon(A^l \cdot, A^l \cdot)](A^{-l}x, A^{-l}y)| \leq C/\rho(x-y) = Cb^{-l} \qquad \text{for } |\alpha|, |\beta| \leq r,$$

*where $l \in \mathbb{Z}$ is the unique integer such that $x - y \in B_{l+1} \setminus B_l$.*

*Furthermore, if $\Phi$ consists of functions with vanishing moments up to order $s < r \ln \lambda_- / \ln \lambda_+$ then $T_\epsilon^*(x^\alpha) = 0$ for all $|\alpha| \leq s$.*

PROOF. For the sake of simplicity we are going to present the calculations only when $\Psi$ and $\Phi$ consist of a single function. The kernel $K_\epsilon$ of the operator $T_\epsilon$ given by (5.7) is given by

$$(5.9) \qquad \begin{aligned} K_\epsilon(x,y) &= \sum_{(j,k) \in \mathbb{Z} \times \mathbb{Z}^n} \epsilon_{j,k} \phi_{j,k}(x) \overline{\psi_{j,k}(y)} \\ &= \sum_{(j,k) \in \mathbb{Z} \times \mathbb{Z}^n} \epsilon_{j,k} b^j \phi(A^j x - k) \overline{\psi(A^j y - k)}. \end{aligned}$$

The formula (5.9) makes perfect sense if all but finitely many of $\epsilon_{j,k}$'s are zeroes. In general, the above series converges absolutely for all $x \neq y$ and for any boundedly supported $f \in L^2(\mathbb{R}^n)$

$$(5.10) \qquad T_\epsilon f(x) = \int_{\mathbb{R}^n} K_\epsilon(x,y) f(y) dy \qquad \text{for } x \notin \overline{\operatorname{supp} f}.$$

Furthermore, we claim that the convergence in (5.9) is uniform on compact subsets of $\Omega = \{(x,y) \in \mathbb{R}^n \times \mathbb{R}^n : x \neq y\}$. Suppose that $\rho(x-y) \geq b^l$, i.e., $x - y \in (B_l)^c$

for some $l \in \mathbb{Z}$. By the Cauchy-Schwarz inequality

$$\sum_{j \leq -l} \sum_{k \in \mathbb{Z}^n} |\epsilon_{j,k} \phi_{j,k}(x) \overline{\psi_{j,k}(y)}| = \sum_{j \leq -l} b^j \sum_{k \in \mathbb{Z}^n} |\phi(A^j x - k) \psi(A^j y - k)|$$

(5.11)
$$\leq \sum_{j \leq -l} b^j \left( \sum_{k \in \mathbb{Z}^n} |\phi(A^j x - k)|^2 \right)^{1/2} \left( \sum_{k \in \mathbb{Z}^n} |\psi(A^j y - k)|^2 \right)^{1/2}$$

$$\leq C \sum_{j \leq -l} b^j = C \frac{b}{b-1} b^{-l}.$$

Here the constant $C$ is such that

$$\sum_{k \in \mathbb{Z}^n} |\psi(z-k)|^2 \leq C \quad \text{and} \quad \sum_{k \in \mathbb{Z}^n} |\phi(z-k)|^2 \leq C \quad \text{for all } z \in \mathbb{R}^n.$$

To estimate the sum over $j > -l$ we use the inequality

(5.12) $\quad |\phi(A^j x - k) \psi(A^j y - k)| \leq C (1 + |A^j x - k|)^{-n-1} (1 + |A^j(x-y)|)^{-N},$

for some $N$ satisfying $b \lambda_-^{-N} < 1$. (5.12) holds since both $\phi$ and $\psi$ decay polynomially fast and

$$(1 + |z|)(1 + |z'|) \geq 1 + |z - z'| \quad \text{for any } z, z' \in \mathbb{R}^n.$$

By (5.12)

(5.13)
$$\sum_{j > -l} \sum_{k \in \mathbb{Z}^n} |\epsilon_{j,k} \phi_{j,k}(x) \overline{\psi_{j,k}(y)}|$$

$$\leq C \sum_{j > -l} b^j \sum_{k \in \mathbb{Z}^n} (1 + |A^j x - k|)^{-n-1} (1 + |A^j(x-y)|)^{-N}$$

$$\leq C \sum_{j > -l} b^j |A^j(x-y)|^{-N} \leq C \sum_{j > -l} b^j (1/c \lambda_-^{j+l})^{-N}$$

$$\leq C b^{-l} \sum_{j > 0} (b \lambda_-^{-N})^j \leq C b^{-l},$$

since $A^j(x - y) \in (B_{j+l})^c$. Combining (5.11) and (5.13) we see that the series in (5.9) converges absolutely and uniformly in the region $\rho(x - y) \geq b^l$. Since $l \in \mathbb{Z}$ is arbitrary, we have uniform convergence on compact subsets of $\Omega$. Furthermore, the estimate (5.8) holds for $\alpha = \beta = 0$. Since (5.10) holds for $\epsilon$'s with all but finitely many nonzero coefficients it holds for any $\epsilon$ by the Lebesgue Dominated Convergence Theorem.

Suppose now that $x - y \in B_{l+1} \setminus B_l$ for some $l \in \mathbb{Z}$. Formally, for any multi-indices $|\alpha|, |\beta| \leq r$ we have

(5.14)
$$\partial_y^\alpha \partial_x^\beta [K(A^l \cdot, A^l \cdot)](A^{-l} x, A^{-l} y)$$
$$= \sum_{(j,k) \in \mathbb{Z} \times \mathbb{Z}^n} \epsilon_{j,k} b^j \partial^\beta [\phi(A^{j+l} \cdot - k)](A^{-l} x) \overline{\partial^\alpha [\psi(A^{j+l} \cdot - k)](A^{-l} y)}.$$

To justify (5.14) we will show that the above series converges absolutely to some value smaller than $C b^{-l}$. As before we estimate separately the sum over $j \leq -l$ and $j > -l$. By the chain rule

$$|\partial^\alpha [\psi(A^{j+l} \cdot - k)](A^{-l} y)| \leq \|A^{j+l}\|^{|\alpha|} \|\mathfrak{D}^{|\alpha|} \psi(A^j y - k)\| \leq C \|\mathfrak{D}^{|\alpha|} \psi(A^j y - k)\|,$$

for $j \leq -l$. Since $\phi$ and $\psi$ are $r$-regular, by the Cauchy-Schwarz inequality

$$\sum_{j \leq -l} b^j \sum_{k \in \mathbb{Z}^n} |\partial^\beta [\phi(A^{j+l} \cdot -k)](A^{-l}x)| |\partial^\alpha [\psi(A^{j+l} \cdot -k)](A^{-l}y)|$$

(5.15)
$$\leq C \sum_{j \leq -l} b^j \left( \sum_{k \in \mathbb{Z}^n} \|\mathfrak{D}^{|\beta|} \phi(A^j x - k)\|^2 \right)^{1/2} \left( \sum_{k \in \mathbb{Z}^n} \|\mathfrak{D}^{|\alpha|} \psi(A^j y - k)\|^2 \right)^{1/2}$$
$$\leq C b^{-l}.$$

To estimate the sum over $j > -l$ we use the inequality

$$|\partial^\beta [\phi(A^{j+l} \cdot -k)](A^{-l}x)| |\partial^\alpha [\psi(A^{j+l} \cdot -k)](A^{-l}y)|$$
$$\leq C^2 \|A^{j+l}\|^{2r} \|\mathfrak{D}^{|\beta|} \phi(A^j x - k)\| \cdot \|\mathfrak{D}^{|\alpha|} \psi(A^j y - k)\|$$
$$\leq C \lambda_+^{2r(j+l)} (1 + |A^j x - k|)^{-n-1} (1 + |A^j (x - y)|)^{-N},$$

for some $N$ satisfying $b \lambda_+^{2r} \lambda_-^{-N} < 1$. The above holds since both $\phi$ and $\psi$ are $r$-regular. Therefore

$$\sum_{j \leq -l} b^j \sum_{k \in \mathbb{Z}^n} |\partial^\beta [\phi(A^{j+l} \cdot -k)](A^{-l}x)| |\partial^\alpha [\psi(A^{j+l} \cdot -k)](A^{-l}y)|$$

(5.16)
$$\leq \sum_{j \leq -l} C \lambda_+^{2r(j+l)} b^j \sum_{k \in \mathbb{Z}^n} (1 + |A^j x - k|)^{-n-1} |A^j (x - y)|^{-N}$$
$$\leq C b^{-l} \sum_{j > 0} (b \lambda_+^{2r} \lambda_-^{-N})^j \leq C b^{-l},$$

since $A^j(x - y) \in B_{j+l+1} \setminus B_{j+l}$. Thus, the series in (5.14) converges absolutely and we have equality there. By (5.15) and (5.16) the estimate (5.8) holds.

Finally, suppose $\phi$ has vanishing moments up to order $s < r \ln \lambda_- / \ln \lambda_+$, i.e., $\int \phi(x) x^\alpha dx = 0$ for $|\alpha| \leq s$. Take any $f \in L^2$ with compact support and vanishing moments up to order $r - 1$. Assume that $\text{supp } f \subset B_l$ for some $l \in \mathbb{Z}$. Since $T_\varepsilon$ are (CZ-$r$) with uniform constants

(5.17)
$$\int_{B_{l+\omega}} |T_\varepsilon f(x)|^2 dx \leq C$$
$$|T_\varepsilon f(x)| \leq C \rho(A^{-l-\omega} x)^{-\delta} \qquad \text{for } x \in (B_{l+\omega})^c$$

by the proof of Lemma 9.5 in Chapter 1, where the constant $C$ is independent of the choice of $\varepsilon$. Here $\delta = r \ln \lambda_- / \ln b + 1$. To guarantee the integrability of $T_\varepsilon f(x)(1 + |x|)^s$ we must have $\delta > s \ln \lambda_+ / \ln b + 1$ and thus $s < r \ln \lambda_- / \ln \lambda_+$. By (5.7) we have

(5.18)
$$\int T_\varepsilon f(x) x^\alpha dx = 0 \qquad \text{for } |\alpha| \leq s,$$

if $\epsilon_{j,k} = 0$ for all but finitely many $(j,k)$'s. Given a general $\epsilon = (\epsilon_{j,k})$ with $|\epsilon_{j,k}| \leq 1$ we define the sequence $(\epsilon^i)_{i \in \mathbb{N}}$ by

$$\epsilon_{j,k}^i = \begin{cases} \epsilon_{j,k} & \text{if } |j|, |k| \leq i, \\ 0 & \text{otherwise}. \end{cases}$$

Since $K_{\epsilon^i}(x,y)$ converges to $K_\epsilon(x,y)$ as $i \to \infty$ uniformly on $\{(x,y): \rho(x-y) \geq b^l\}$ we have
$$T_{\epsilon^i}f(x) \to T_\epsilon f(x) \quad \text{as } i \to \infty \quad \text{for } x \in (B_{l+\omega})^c.$$
Since $\int T_{\epsilon^i}f(x)x^\alpha dx = 0$ for all $|\alpha| \leq s$, $i \in \mathbb{N}$ we have (5.18) by the Lebesgue Dominated Convergence Theorem and (5.17). This ends the proof of Lemma 5.5. □

THEOREM 5.6. *Suppose $\Psi$ and $\Phi$ are $r$-regular Bessel multiwavelets for some $r$. Suppose also that $p$ satisfies*

(5.19) $$0 \leq 1/p - 1 < \frac{(\ln \lambda_-)^2}{\ln b \ln \lambda_+} r.$$

*If $\Phi$ consists of functions with vanishing moments up to order $s = \lfloor(1/p - 1)\ln b/\ln \lambda_-\rfloor$ then for any sequence of scalars $\epsilon = (\epsilon^l_{j,k})_{(l,j,k)\in\Lambda}$ with $|\epsilon^l_{j,k}| \leq 1$ then the operator $T_\epsilon$ given by (5.7) extends to a bounded operator from $H^p_A(\mathbb{R}^n)$ into $H^p_A(\mathbb{R}^n)$ with the norm independent of $\epsilon$.*

PROOF. The proof is an immediate consequence of Lemma 5.5 and Theorem 9.8 in Chapter 1 since $s < r\ln\lambda_-/\ln\lambda_+$. □

We shall need one more result.

LEMMA 5.7. *With all the assumptions of Theorem 5.6 we also assume that we have a sequence $(\epsilon^i)_{i\in\mathbb{N}}$ in the unit ball of $\ell^\infty(\Lambda)$, i.e., $|\epsilon^{i,l}_{j,k}| \leq 1$ for $i \in \mathbb{N}$, $(l,j,k) \in \Lambda$ such that $\epsilon^i \to \epsilon$ weak-$*$ in $\ell^\infty$ as $i \to \infty$. In other words,*

(5.20) $$\epsilon^{i,l}_{j,k} \to \epsilon^l_{j,k} \quad \text{as } i \to \infty \quad \text{for all } (l,j,k) \in \Lambda.$$

*Then for any $f \in H^p_A(\mathbb{R}^n)$*

(5.21) $$T_{\epsilon^i}f \to T_\epsilon f \quad \text{in } H^p_A \quad \text{as } i \to \infty.$$

PROOF. Suppose first that $f \in L^2$ is compactly supported with vanishing moments up to order $s$, say $\operatorname{supp} f \subset B_l$ for some $l \in \mathbb{Z}$. By (5.7), (5.17), and (5.20)

(5.22) $$\int_{B_{l+\omega}} |(T_{\epsilon^i} - T_\epsilon)f(x)|^2 dx \to 0 \quad \text{as } i \to \infty,$$
$$|T_{\epsilon^i}f(x)|, |T_\epsilon f(x)| \leq C\rho(A^{-l-\omega}x)^{-\delta} \quad \text{for } x \in (B_{l+\omega})^c,$$
$$|(T_{\epsilon^i} - T_\epsilon)f(x)| \to 0 \quad \text{as } i \to \infty \quad \text{for } x \in (B_{l+\omega})^c.$$

Furthermore $\int (T_{\epsilon^i} - T_\epsilon)f(x)x^\alpha dx = 0$ for $|\alpha| \leq s$. To show that

(5.23) $$\|(T_{\epsilon^i} - T_\epsilon)f\|_{H^p_{2,s}} \to 0 \quad \text{as } i \to \infty,$$

a slight modification of the proof of Lemma 9.3 in Chapter 1 is required. The proof proceeds in the same manner, except that we are going to show that $g_{j+1} - g_j$ are appropriate multiples of $(p,2,s)$-atoms. We not only have
$$\|g_{j+1} - g_j\|_2 \leq C_4 b^{(j+1)(1/2-1/p)} b^{(\delta-1/p)(l+\omega-j)} \quad \text{for } j \geq l+\omega,$$

where $g_j = \tilde{\pi}_{B_j}((T_{\epsilon^i} - T_\epsilon)f)\mathbf{1}_{B_j}$, but also $\|g_{j+1} - g_j\|_2 \to 0$ as $i \to \infty$ by (5.22) and the Lebesgue Dominated Convergence Theorem. This shows (5.23) for all $(p, 2, s)$-atoms $f$. In particular, for any $(p, 2, s)$-atom $a$ we have

$$(5.24) \qquad T_\epsilon(a) = \sum_{(l,j,k) \in \Lambda} \epsilon^l_{j,k} \langle a, \psi^l_{j,k} \rangle \phi^l_{j,k},$$

with the unconditional convergence in $H^p_A$. Given a general element $f \in H^p_A$ we find its atomic decomposition $f = \sum_{j \in \mathbb{N}} \kappa_j a_j$ with $\sum_{j \in \mathbb{N}} |\kappa_j|^p < \infty$ and $a_j$'s are $(p, 2, s)$-atoms. By (5.23) and the uniform boundedness of $T_{\epsilon^i}$'s

$$T_{\epsilon^i} f = \sum_{j \in \mathbb{N}} \kappa_j T_{\epsilon^i} a_j \to \sum_{j \in \mathbb{N}} \kappa_j T_\epsilon a_j = T_\epsilon f \quad \text{as } i \to \infty.$$

Therefore (5.21) holds. This ends the proof of Lemma 5.7. $\square$

As a corollary of the above considerations we have for all $f \in H^p_A$

$$(5.25) \qquad \begin{aligned} T_\epsilon(f) &= \sum_{i \in \mathbb{N}} \kappa_i T_\epsilon a_i = \sum_{i \in \mathbb{N}} \kappa_i \sum_{(l,j,k) \in \Lambda} \epsilon^l_{j,k} \langle a_i, \psi^l_{j,k} \rangle \phi^l_{j,k} \\ &= \sum_{(l,j,k) \in \Lambda} \epsilon^l_{j,k} \left( \sum_{i \in \mathbb{N}} \kappa_i \langle a_i, \psi^l_{j,k} \rangle \right) \phi^l_{j,k} = \sum_{(l,j,k) \in \Lambda} \epsilon^l_{j,k} \langle f, \psi^l_{j,k} \rangle \phi^l_{j,k}, \end{aligned}$$

since the above series converges unconditionally and the $\psi^l_{j,k}$'s belong to the Campanato space $C^{1/p-1}_{\infty,s}$. The formula (5.25) is of practical importance because it says that (5.7) holds also if $f \in H^p_A$ for an appropriate range of $p$'s.

**Wavelets as an unconditional basis.** We are now ready to harvest the fruit of our labor.

THEOREM 5.8. *Suppose $\Psi$ is an $r$-regular multiwavelet consisting of functions with vanishing moments up to the order $\lfloor r \ln \lambda_- / \ln \lambda_+ \rfloor$. Then $(\psi^l_{j,k})_{(l,j,k) \in \Lambda}$ forms an unconditional basis for $H^p_A$ for all $p$ satisfying (5.19).*

PROOF. We can apply Theorem 5.6 and Lemma 5.7, since

$$s = \lfloor (1/p - 1) \ln b / \ln \lambda_- \rfloor \le \lfloor r \ln \lambda_- / \ln \lambda_+ \rfloor.$$

By Lemma 5.7 and (5.25)

$$f = \sum_{(l,j,k) \in \Lambda} \langle f, \psi^l_{j,k} \rangle \psi^l_{j,k},$$

and the convergence is unconditional in $H^p_A$. Therefore, the system $(\psi^l_{j,k})_{(l,j,k) \in \Lambda}$ is a basis for $H^p_A$, since $\Psi$ is an orthonormal multiwavelet. $\square$

We remark that the assumption in Theorem 5.8 that $\Psi$ consists of functions with vanishing moments is, in fact, a consequence of $\Psi$ being an $r$-regular multiwavelet. This is a well-known fact for dyadic wavelets which can be found in [Bt, Da2, Me]. This result was extended to general dilations with integer entries by the author [Bo4, Theorem 4].

THEOREM 5.9. *Suppose $A$ is a dilation with integer entries and $\Psi$ is an $r$-regular orthonormal multiwavelet for some $r \in \mathbb{N}$. Then*

$$\int_{\mathbb{R}^n} x^\alpha \psi^l(x) dx = 0 \qquad \text{for all } l = 1, \ldots, L, \ |\alpha| < r \ln \lambda_- / \ln \lambda_+.$$

Therefore, the assumption in Theorem 5.8 that $\Psi$ has vanishing moments is an automatic consequence of $r$-regularity of $\Psi$. Even though the proof of Theorem 5.9 in [Bo4] works only for integer dilations, it is very likely that Theorem 5.9 holds for all dilations. For example, Theorem 5.9 holds trivially for dilations $A$ satisfying (3.11), since such dilations do not allow $r$-regular wavelets. Hence, we conjecture that Theorem 5.9 is valid for all dilations.

By Theorem 5.8 and standard interpolation theory we can also conclude that a 1-regular multiwavelet generates an unconditional basis in $L^p$ for $1 < p < \infty$. This can be shown to hold under fairly minimal decay conditions on $\Psi$, see [Po, Wo4].

What can we say if for some dilation $A$ we cannot find an orthonormal $r$-regular multiwavelets, e.g. when $A$ does not preserve any lattice? Even though Theorem 5.8 still holds, it may become a vacuous statement due to the lack of $r$-regular wavelets, see Theorem 3.1. Nevertheless, by Section 4 we can always find a non-orthonormal wavelet $\psi \in \mathcal{S}$ such that $(\psi_{j,k})_{j \in \mathbb{Z}, k \in \mathbb{Z}^n}$ forms a tight frame in $L^2(\mathbb{R}^n)$ and $\psi$ has all vanishing moments. We can now show that this wavelet forms a "tight frame" for $H_A^p$.

THEOREM 5.10. *Suppose $\Psi$ is an $r$-regular tight frame multiwavelet and $p$ satisfies (5.19). If $\Psi$ consists of functions with vanishing moments up to the order $s = \lfloor (1/p - 1) \ln b / \ln \lambda_- \rfloor$, then $(\psi_{j,k}^l)_{(l,j,k) \in \Lambda}$ forms a tight frame for $H_A^p$, i.e.,*

$$(5.26) \qquad f = \sum_{(l,j,k) \in \Lambda} \langle f, \psi_{j,k}^l \rangle \psi_{j,k}^l \qquad \text{for all } f \in H_A^p,$$

*and the convergence is unconditional in $H_A^p$.*

The proof follows verbatim the proof of Theorem 5.8.

## 6. Characterization of $H^p$ in terms of wavelet coefficients

In this section we will characterize the sequence space of wavelet expansion coefficients $(\langle f, \psi_{j,k}^l \rangle)$ for $f \in H_A^p$, where $\Lambda = \{(l, j, k) : l = 1, \ldots, L, j \in \mathbb{Z}, k \in \mathbb{Z}^n\}$.

Let $(\epsilon_{j,k}^l)_{(l,j,k) \in \Lambda}$ be the sequence of *independent identically distributed* (i.i.d.) random Bernoulli variables with $P(\epsilon_{j,k}^l = 1) = P(\epsilon_{j,k}^l = -1) = 1/2$. In other words, if we consider the probability space $([0, 1], dt)$ then we can think of $\epsilon_{j,k}^l(t) = r_{\sigma(l,j,k)}(t)$, where $\sigma : \Lambda \to \mathbb{N}$ is a bijection and $(r_i(t))_{i \in \mathbb{N}}$ are the *Rademacher functions* defined by $r_i(t) = \operatorname{sign}(\sin(2^i \pi t))$ for $t \in [0, 1]$.

The fundamental result concerning the means of i.i.d. random Bernoulli variables is Khinchin's (Хинчин) inequality.

THEOREM 6.1. *For any $0 < p < \infty$ there are positive constants $A_p$, $B_p$ such that*

$$(6.1) \quad A_p \left( \sum_{i \in \mathbb{N}} |c_i|^2 \right)^{1/2} \leq \left( \int_0^1 \left| \sum_{i \in \mathbb{N}} c_i r_i(t) \right|^p dt \right)^{1/p} \leq B_p \left( \sum_{i \in \mathbb{N}} |c_i|^2 \right)^{1/2},$$

*for any sequence of scalars $(c_i)$.*

The inequality (6.1) means that $\sum_i c_i r_i$ converges unconditionally in $L^p$ if and only if $\sum_i c_i r_i$ converges in $L^2$, i.e., $\sum_i |c_i|^2 < \infty$. It diverges in the $L^p$ norm to infinity if $\sum_i |c_i|^2 = \infty$.

We shall also need the Fefferman-Stein vector-valued maximal inequality, for a proof see [FS1, St2]. Here $M$ is the Hardy-Littlewood maximal operator on a space of homogeneous type.

THEOREM 6.2. *Suppose $1 < p < \infty$ and $1 < q \leq \infty$. Then there is a constant $C_{p,q}$ depending on $p$ and $q$ such that*

$$(6.2) \quad \left\| \left( \sum_{i \in \mathbb{N}} |M f_i|^q \right)^{1/q} \right\|_{L^p} \leq C_{p,q} \left\| \left( \sum_{i \in \mathbb{N}} |f_i|^q \right)^{1/q} \right\|_{L^p}.$$

Our goal is to show the following lemma.

LEMMA 6.3. *Suppose $0 < p \leq 1$ and $\Psi$ is an 0-regular Bessel multiwavelet consisting of nonzero functions with vanishing moments up to the order $s = \lfloor (1/p - 1) \ln b / \ln \lambda_- \rfloor$. Given a sequence of scalars $(c_{j,k}^l)_{(l,j,k) \in \Lambda}$, the following are equivalent:*

$$(6.3) \quad \sum_{(l,j,k) \in \Lambda} c_{j,k}^l \psi_{j,k}^l \quad \text{converges unconditionally in } H_A^p,$$

$$(6.4) \quad \left( \sum_{(l,j,k) \in \Lambda} |c_{j,k}^l|^2 |\psi_{j,k}^l(x)|^2 \right)^{1/2} \in L^p(\mathbb{R}^n),$$

$$(6.5) \quad \left( \sum_{(l,j,k) \in \Lambda} |c_{j,k}^l|^2 |(\mathbf{1}_{E_l})_{j,k}(x)|^2 \right)^{1/2} \in L^p(\mathbb{R}^n),$$

*for any (or some) bounded measurable sets $E_l \subset \mathbb{R}^n$ with $|E_l| > 0$, $l = 1, \ldots, L$.*

Before proceeding with the proof of Lemma 6.3 we need to introduce some terminology. Let $\mathcal{I}$ be the family of dilated cubes, i.e.,
$$\mathcal{I} = \{A^j([0,1]^n + k) : j \in \mathbb{Z}, k \in \mathbb{Z}^n\}.$$
Given $I = A^j([0,1]^n + k) \in \mathcal{I}$ we define the *scale* of $I$ by $\mathrm{scale}(I) = j$. Let $g$ be the square function given by (6.5) for $E_l = [0,1]^n$,

$$(6.6) \quad g(x) = \left( \sum_{(l,j,k) \in \Lambda} |c_{j,k}^l|^2 |(\mathbf{1}_{E_l})_{j,k}(x)|^2 \right)^{1/2} = \left( \sum_{l=1}^{L} \sum_{I \in \mathcal{I}} |c_I^l|^2 |I|^{-1} \mathbf{1}_I(x) \right)^{1/2},$$

where we use the convention $c_I^l = c_{-j,k}^l$ for a cube $I = A^j([0,1]^n + k)$, $j \in \mathbb{Z}$, $k \in \mathbb{Z}^n$.

DEFINITION 6.4. Suppose $\mathcal{I}' \subset \mathcal{I}$. We say that the cube $I$ is *stacked below* the cube $J$ within the family $\mathcal{I}'$, and write $I \preccurlyeq_{\mathcal{I}'} J$, if there is a chain of cubes $I = I_0, I_1, \ldots, I_s = J \in \mathcal{I}'$ such that

$$\mathrm{scale}(I_i) < \mathrm{scale}(I_{i+1}) \quad \text{and} \quad |I_i \cap I_{i+1}| \neq 0 \quad \text{for all } i = 0, \ldots, s-1.$$

The relation $\preccurlyeq_{\mathcal{I}'}$ induces a partial order in $\mathcal{I}'$. Let $\max(\mathcal{I}')$ be the set of maximal elements in $\mathcal{I}'$ with respect to $\preccurlyeq_{\mathcal{I}'}$.

If a subfamily $\mathcal{I}'$ does not contain arbitrary large cubes, i.e., $\sup_{I \in \mathcal{I}'} \mathrm{scale}(I) < \infty$, then for any cube $I \in \mathcal{I}'$ there is always a cube $J \in \max(\mathcal{I}')$ with $I \preccurlyeq_{\mathcal{I}'} J$. In general, a maximal cube is not unique unless, for example, the dilation $A = 2Id$ and we work with nicely nested dyadic cubes. We shall need a simple lemma.

LEMMA 6.5. *There is a universal constant $\eta \in \mathbb{N}$ such that whenever we have two cubes $I, J \in \mathcal{I}'$ with $I \preccurlyeq_{\mathcal{I}'} J = A^j([0,1]^n + k)$ then $I \subset A^j(B_\eta + k)$.*

In the above lemma a family $\mathcal{I}'$ does not play any role and can be substituted by the whole family $\mathcal{I}$.

PROOF. Suppose first that $J = [0,1]^n + k$ for some $k \in \mathbb{Z}$. For any integer the diameter of a cube with $\mathrm{scale}(I) = j$ is at most

$$\mathrm{diam}(I) = \mathrm{diam}(A^j[0,1]^n) \leq \|A^j\| \mathrm{diam}([0,1]^n) = 2^n \|A^j\|.$$

Whenever we have a chain of cubes $I_0, I_1, \ldots, I_s = J \in \mathcal{I}$ satisfying Definition 6.4 then

$$\mathrm{diam}\left( \bigcup_{i=0}^{s} I_i \right) \leq \sum_{i=0}^{s} \mathrm{diam}(I_i) \leq 2^n \sum_{j=-\infty}^{0} \|A^j\| < \infty.$$

Therefore, there exists $\eta \in \mathbb{N}$ such that $I \subset B_\eta + k$ whenever $I \preccurlyeq_{\mathcal{I}} J$. If $J = A^j([0,1]^n + k)$ is arbitrary then $I \preccurlyeq_{\mathcal{I}} J \iff A^{-j}I \preccurlyeq_{\mathcal{I}} A^{-j}J$. Thus, $A^{-j}I \subset B_\eta + k$ and hence $I \subset A^j(B_\eta + k)$. □

We also need the following elementary lemma.

LEMMA 6.6. *Suppose $r_1, r_2 > 0$ and $P$ is an $n \times n$ real matrix with $|Px| \geq r_1|x|$ for all $x \in \mathbb{R}^n$. Then there is a constant $c = c(r_1, r_2)$ depending only on $r_1$ and $r_2$ such that*

$$\#(\mathbb{Z}^n \cap P\mathbf{B}(z, r_2)) \leq c|\det P| \quad \text{for any } z \in \mathbb{R}^n.$$

PROOF. Note that

$$\begin{aligned}
\#(\mathbb{Z}^n \cap P\mathbf{B}(z,r_2)) &= \#(P^{-1}\mathbb{Z}^n \cap \mathbf{B}(z,r_2)) \\
&\leq \#\{k \in \mathbb{Z}^n : P^{-1}(k+(0,1)^n) \cap \mathbf{B}(0,r_2) \neq \emptyset\} \\
&\leq \#\{k \in \mathbb{Z}^n : P^{-1}(k+(0,1)^n) \subset \mathbf{B}(0,r_2+2^n/r_1)\} \\
&\leq \frac{|\mathbf{B}(0,r_2+2^n/r_1)|}{|P^{-1}((0,1)^n)|} = |\mathbf{B}(0,r_2+2^n/r_1)| \cdot |\det P|,
\end{aligned}$$

since $\operatorname{diam} P^{-1}((0,1)^n) \leq 2^n/r_1$ and the family $\{P^{-1}(k+(0,1)^n) : k \in \mathbb{Z}^n\}$ consists of pairwise disjoint sets. $\square$

PROOF OF LEMMA 6.3. Assume that (6.3) holds. Let $(\epsilon_{j,k}^l)$ be a sequence i.i.d. random Bernoulli variables. By Lemma 5.4 there exists a constant $C$ such that for any subset $F \subset \Lambda$

$$(6.7) \qquad \left\| \sum_{(l,j,k) \in F} \epsilon_{j,k}^l(t) c_{j,k}^l \psi_{j,k}^l \right\|_{H^p} \leq C \qquad \text{for any } t \in [0,1].$$

The identity on $L^2(\mathbb{R}^n)$ is a Calderón-Zygmund operator (of any order) hence it extends to a bounded operator from $H^p$ to $L^p$ with norm 1 by Theorem 9.9 in Chapter 1. If $F \subset \Lambda$ is finite then by Khinchin's inequality, the Fubini Theorem and (6.7)

$$A_p \int_{\mathbb{R}^n} \left( \sum_{(l,j,k) \in F} |c_{j,k}^l|^2 |\psi_{j,k}^l(x)|^2 \right)^{p/2} dx \leq \int_{\mathbb{R}^n} \int_0^1 \left| \sum_{(l,j,k) \in F} \epsilon_{j,k}^l(t) c_{j,k}^l \psi_{j,k}^l(x) \right|^p dt\, dx$$

$$\leq \int_0^1 \left\| \sum_{(l,j,k) \in F} \epsilon_{j,k}^l(t) c_{j,k}^l \psi_{j,k}^l \right\|_{L^p}^p dt \leq C^p.$$

Since $F \subset \Lambda$ is arbitrary, by Fatou's Lemma we have

$$\int_{\mathbb{R}^n} \left( \sum_{(l,j,k) \in \Lambda} |c_{j,k}^l|^2 |\psi_{j,k}^l(x)|^2 \right)^{p/2} dx \leq C^p/A_p,$$

which shows (6.4).

Assume now that (6.4) holds. Let $\delta > 0$ be sufficiently small so that $\tilde{E}_l = \{x \in \mathbb{R}^n : |\psi^l(x)| > \delta\}$ has nonzero measure for every $l = 1, \ldots, L$. Since

$$\sum_{(l,j,k) \in \Lambda} |c_{j,k}^l|^2 |\psi_{j,k}^l(x)|^2 \geq \delta^2 \sum_{(l,j,k) \in \Lambda} |c_{j,k}^l|^2 |(\mathbf{1}_{\tilde{E}_l})_{j,k}(x)|^2 \qquad \text{for all } x \in \mathbb{R}^n,$$

(6.5) holds for some $\tilde{E}_l$'s. Choose any other bounded $E_l$'s with nonzero measure. We can then find $\delta' > 0$ so that $M(\mathbf{1}_{\tilde{E}_l})(x) \geq \delta' \mathbf{1}_{E_l}(x)$ for all $x \in \mathbb{R}^n$, where $M$ is the Hardy-Littlewood maximal operator associated to the dilation $A$ given by (3.15) in Chapter 1. Since $M$ commutes with translations and dilations by the power of $A$ we have $(M(\mathbf{1}_{\tilde{E}_l}))_{j,k}(x) = M((\mathbf{1}_{\tilde{E}_l})_{j,k}^r)^{1/r}(x)$ for any $r > 0$. Hence, for

$0 < r < p$ by Lemma 6.2

$$(\delta')^p \int_{\mathbb{R}^n} \left( \sum_{(l,j,k) \in \Lambda} |c_{j,k}^l|^2 |(\mathbf{1}_{E_l})_{j,k}(x)|^2 \right)^{p/2} dx$$

$$\leq \int_{\mathbb{R}^n} \left( \sum_{(l,j,k) \in \Lambda} |c_{j,k}^l|^2 M((\mathbf{1}_{\tilde{E}_l})_{j,k}^r)^{2/r}(x) \right)^{p/2} dx$$

$$= \int_{\mathbb{R}^n} \left( \sum_{(l,j,k) \in \Lambda} |M(|c_{j,k}^l|^r (\mathbf{1}_{\tilde{E}_l})_{j,k}^r)(x)|^{2/r} \right)^{(r/2)(p/r)} dx$$

$$\leq (C_{p/r,2/r})^{p/r} \int_{\mathbb{R}^n} \left( \sum_{(l,j,k) \in \Lambda} |c_{j,k}^l|^2 |(\mathbf{1}_{\tilde{E}_l})_{j,k}(x)|^2 \right)^{p/2} dx,$$

which shows (6.5).

Finally, assume (6.5) holds with $E_l = [0,1]^n$ for $l = 1, \ldots, L$. We shall show that after some rearrangement the series in (6.3) is an appropriate combination of molecules.

For each $r \in \mathbb{Z}$ define $\Omega_r = \{x \in \mathbb{R}^n : g(x) > 2^r\}$, where $g(x)$ is given by (6.6). Clearly, $\Omega_{r+1} \subset \Omega_r$ and

(6.8)
$$\sum_{r \in \mathbb{Z}} 2^{pr} |\Omega_r| = \sum_{r \in \mathbb{Z}} 2^{pr} \sum_{s=r}^{\infty} |\Omega_{s+1} \setminus \Omega_s| = \sum_{s \in \mathbb{Z}} \sum_{r=-\infty}^{s} 2^{pr} |\Omega_{s+1} \setminus \Omega_s|$$
$$= \frac{1}{1-2^{-p}} \sum_{s \in \mathbb{Z}} 2^{ps} |\Omega_{s+1} \setminus \Omega_s| \leq \frac{1}{1-2^{-p}} \int_{\mathbb{R}^n} |g(x)|^p dx.$$

We also define a subfamily of dilated cubes

$$\mathcal{I}_r = \{I \in \mathcal{I} : |I \cap \Omega_r| \geq |I|/2 \text{ and } |I \cap \Omega_{r+1}| < |I|/2\}.$$

Note that for any $I \in \mathcal{I}$ with $c_I^l \neq 0$ for some $l = 1, \ldots, L$, there is a unique $r \in \mathbb{Z}$ such that $I \in \mathcal{I}_r$.

For each $r \in \mathbb{Z}$, we have

$$\mathcal{I}_r = \bigcup_{J \in \max(\mathcal{I}_r)} \{I \in \mathcal{I}_r : I \preccurlyeq_{\mathcal{I}_r} J\},$$

since the subfamily $\mathcal{I}_r$ does not contain arbitrary large cubes by $|\Omega_r| < \infty$. By induction on $J \in \max(\mathcal{I}_r)$ we can find pairwise disjoint subfamilies $\mathcal{I}_{r,J}$ satisfying

(6.9)
$$\mathcal{I}_{r,J} \subset \{I \in \mathcal{I}_r : I \preccurlyeq_{\mathcal{I}_r} J\}, \quad \mathcal{I}_r = \bigcup_{J \in \max(\mathcal{I}_r)} \mathcal{I}_{r,J},$$
$$\mathcal{I}_{r,J} \cap \mathcal{I}_{r,J'} = \emptyset \quad \text{for } J \neq J' \in \max(\mathcal{I}_r).$$

We are now ready to rearrange the formal series in (6.3) as a combination of molecules. Using the convention $c_I^l = c_{-j,k}^l$, $\psi_I = \psi_{-j,k}$ for a cube $I = A^j([0,1]^n + k)$ we can write

$$\sum_{(l,j,k) \in \Lambda} c_{j,k}^l \psi_{j,k}^l = \sum_{l=1}^{L} \sum_{I \in \mathcal{I}} c_I^l \psi_I^l = \sum_{r \in \mathbb{Z}} \sum_{J \in \max(\mathcal{I}_r)} \left( \sum_{l=1}^{L} \sum_{I \in \mathcal{I}_{r,J}} c_I^l \psi_I^l \right),$$

where $\mathcal{I}_r$ and $\mathcal{I}_{r,j}$ are given by (6.9). To show that this series converges in $H_A^p$ it suffices to show that $\sum_{I \in \mathcal{I}_{r,J}} c_I^l \psi_I^l$ is an element of $H_A^p$ and

$$\text{(6.10)} \qquad \sum_{r \in \mathbb{Z}} \sum_{J \in \max(\mathcal{I}_r)} \left\| \sum_{l=1}^L \sum_{I \in \mathcal{I}_{r,J}} c_I^l \psi_I^l \right\|_{H^p}^p < \infty,$$

by the completeness of $H_A^p$ and $p$-subadditivity of $\|\cdot\|_{H^p}$.

Fix $r \in \mathbb{Z}$ and $J = A^{j_0}([0,1]^n + k_0) \in \max(\mathcal{I}_r)$. We claim that there is a constant $C$ (independent of $r$ and $J$) such that

$$\text{(6.11)} \qquad \left\| \sum_{l=1}^L \sum_{I \in \mathcal{I}_{r,J}} c_I^l \psi_I^l \right\|_{H^p} \leq C 2^r |J|^{1/p}.$$

By Lemma 6.5 we have

$$\text{(6.12)} \quad \begin{aligned} \sum_{l=1}^L \sum_{I \in \mathcal{I}_{r,J}} |c_I^l|^2 &\leq 2 \sum_{l=1}^L \sum_{I \in \mathcal{I}_{r,J}} |c_I^l|^2 |I|^{-1} |I \setminus \Omega_{r+1}| \\ &\leq 2 \sum_{l=1}^L \sum_{I \in \mathcal{I}_{r,J}} \int_{(A^{j_0}(B_\eta + k_0)) \setminus \Omega_{r+1}} |c_I^l|^2 |I|^{-1} \mathbf{1}_I(x) dx \\ &\leq 2 \int_{(A^{j_0}(B_\eta + k_0)) \setminus \Omega_{r+1}} |g(x)|^2 dx \leq 2 b^{j_0 + \eta} 2^{2(r+1)} \\ &= 8 b^{j_0 + \eta} 2^{2r}, \end{aligned}$$

since outside of $\Omega_{r+1}$ we have $g(x) \leq 2^{r+1}$.

Since $\{\psi_I^l\}$ is a Bessel family by (6.12) we have

$$\frac{1}{|B_{j_0+\eta+\omega}|} \int_{B_{j_0+\eta+\omega} + A^{j_0} k_0} \left| \sum_{l=1}^L \sum_{I \in \mathcal{I}_{r,J}} c_I^l \psi_I^l(x) \right|^2 dx$$

$$\leq \frac{1}{b^{j_0+\eta+\omega}} \left\| \sum_{l=1}^L \sum_{I \in \mathcal{I}_{r,J}} c_I^l \psi_I^l \right\|_{L^2}^2 \leq \frac{C}{b^{j_0+\eta+\omega}} \sum_{l=1}^L \sum_{I \in \mathcal{I}_{r,J}} |c_I^l|^2 \leq C 2^{2r}.$$

For any $I \in \mathcal{I}_{r,J}$ by Lemma 6.5 we have $I = A^j([0,1]^n + k) \subset A^{j_0}(B_\eta + k_0)$ and thus $A^j k \in A^{j_0} k_0 + B_{j_0+\eta}$. Therefore, for fixed $j < j_0$ the number of such $k$'s is bounded by Lemma 6.6

$$\text{(6.13)} \qquad \#(\mathbb{Z}^n \cap (A^{j_0-j} k_0 + B_{j_0-j+\eta})) \leq c b^{j_0-j+\eta}.$$

Also for $x \in A^{j_0} k_0 + (B_{j_0+\eta+\omega})^c$ we have

$$\rho(x - A^{j_0} k_0) \leq b^\omega \rho(x - A^j k),$$

since $A^{j_0} k_0 - A^j k \in B_{j_0+\eta}$. Since $\Psi$ is 0-regular, for any $\delta > 0$ there is $C = C(\delta) > 0$ such that $|\psi^l(x)| \leq C \rho(x)^{-\delta}$ for all $x \in \mathbb{R}^n$. In particular, we can take $\delta$ satisfying the hypothesis of Lemma 9.3 in Chapter 1. Therefore, for $x \in A^{j_0} k_0 + (B_{j_0+\eta+\omega})^c$,

$$\begin{aligned} |\psi_I^l(x)|^2 &= |\psi_{-j,k}^l(x)|^2 = b^{-j} |\psi^l(A^{-j} x - k)|^2 \leq C b^{-j} \rho(A^{-j} x - k)^{-2\delta} \\ &\leq C b^{2\delta\omega} b^{-j} \rho(A^{-j}(x - A^{j_0} k_0))^{-2\delta} \\ &= C b^{-j} b^{-2\delta(j_0-j+\eta)} \rho(A^{-j_0-\eta-\omega}(x - A^{j_0} k_0))^{-2\delta}. \end{aligned}$$

Combining the above with (6.12) and (6.13), by the Cauchy-Schwarz inequality we have

$$\Big|\sum_{l=1}^{L}\sum_{I\in\mathcal{I}_{r,J}}c_I^l\psi_I^l(x)\Big|\le\Big(\sum_{l=1}^{L}\sum_{I\in\mathcal{I}_{r,J}}|c_I^l|^2\Big)^{1/2}\Big(\sum_{l=1}^{L}\sum_{I\in\mathcal{I}_{r,J}}|\psi_I^l(x)|^2\Big)^{1/2}$$

$$\le b^{(j_0+\eta)/2}2^r\Big(\sum_{j<j_0}cb^{j_0-j+\eta}Cb^{-j}b^{-2\delta(j_0-j+\eta)}\rho(A^{-j_0-\eta-\omega}(x-A^{j_0}k_0))^{-2\delta}\Big)^{1/2}$$

$$\le C2^r\Big(\sum_{j<j_0}b^{(2-2\delta)(j_0-j)}\Big)^{1/2}\rho(A^{-j_0-\eta-\omega}(x-A^{j_0}k_0))^{-\delta},$$

for $x\in A^{j_0}k_0+(B_{j_0+\eta+\omega})^c$. By the Lebesgue Dominated Convergence Theorem we also have

$$\int_{\mathbb{R}^n}\Big(\sum_{l=1}^{L}\sum_{I\in\mathcal{I}_{r,J}}c_I^l\psi_I^l(x)\Big)x^\alpha dx=0\quad\text{for }|\alpha|\le s.$$

Therefore, all the hypotheses of Lemma 9.3 in Chapter 1 are satisfied and thus,

$$\Big\|\sum_{l=1}^{L}\sum_{I\in\mathcal{I}_{r,J}}c_I^l\psi_I^l\Big\|_{H^p}\le C2^r|B_{j_0+\eta+\omega}|^{1/p}\le C2^r|J|^{1/p}.$$

This proves (6.11). To show (6.10) we use $|J|\le 2|J\cap\Omega_r|$ for $J\in\max(\mathcal{I}_r)$ to obtain

$$\sum_{r\in\mathbb{Z}}\sum_{J\in\max(\mathcal{I}_r)}\Big\|\sum_{l=1}^{L}\sum_{I\in\mathcal{I}_{r,J}}c_I^l\psi_I^l\Big\|_{H^p}^p\le 2C^p\sum_{r\in\mathbb{Z}}\sum_{J\in\max(\mathcal{I}_r)}2^{pr}|J\cap\Omega_r|$$

$$\le 2C^p\sum_{r\in\mathbb{Z}}2^{pr}|\Omega_r|\le\frac{2C^p}{1-2^{-p}}\|g\|_{L^p}^p.$$

Since the above estimate depends only on a magnitude of coefficients $c_I^l$'s, we obtain the unconditional convergence in (6.3). This ends the proof of Lemma 6.3. □

Note that the assumption that $\Psi$ is 0-regular in Lemma 6.3 can be slightly relaxed. It follows from the proof that it suffices to assume $|\psi^l(x)|\le C\rho(x)^{-\delta}$ for $l=1,\ldots,L$ with $\delta>\max(1/p,s\ln\lambda_+/\ln b+1)$ in order to apply Lemma 9.3 in Chapter 1.

We are now ready to prove the characterization of the sequence space of wavelet expansion coefficients of elements in $H_A^p$.

THEOREM 6.7. *Suppose $p$ satisfies (5.19), and $\Psi$ is an $r$-regular tight frame multiwavelet with vanishing moments up to order $s=\lfloor(1/p-1)\ln b/\ln\lambda_-\rfloor$. Then for any $f\in H_A^p$ the series $\sum_{(l,j,k)\in\Lambda}\langle f,\psi_{j,k}^l\rangle\psi_{j,k}^l$ converges unconditionally to $f$ in $H_A^p$, and*

(6.14)
$$\|f\|_{H_A^p}\sim\Big\|\Big(\sum_{(l,j,k)\in\Lambda}|\langle f,\psi_{j,k}^l\rangle|^2|\psi_{j,k}^l(\cdot)|^2\Big)^{1/2}\Big\|_{L^p}$$
$$\sim\Big\|\Big(\sum_{(l,j,k)\in\Lambda}|\langle f,\psi_{j,k}^l\rangle|^2|(\mathbf{1}_{E_l})_{j,k}(\cdot)|^2\Big)^{1/2}\Big\|_{L^p},$$

for any (or some) bounded measurable sets $E_l \subset \mathbb{R}^n$ with $|E_l| > 0$, $l = 1, \ldots, L$. The equivalence constants in (6.14) do not depend on $f$.

If, in addition, $\Psi$ is an orthonormal multiwavelet then for any $(c_{j,k}^l)_{(l,j,k)\in\Lambda}$ satisfying

$$(6.4) \qquad \left( \sum_{(l,j,k)\in\Lambda} |c_{j,k}^l|^2 |\psi_{j,k}^l(x)|^2 \right)^{1/2} \in L^p(\mathbb{R}^n),$$

there is a unique $f \in H_A^p$ such that $c_{j,k}^l = \langle f, \psi_{j,k}^l \rangle$ for all $(l, j, k) \in \Lambda$ and (6.14) holds.

PROOF. By Theorem 5.10 we have (5.26). By Lemma 6.3 we have (6.14). The fact that the equivalence constants in (6.14) do not depend on the choice of $f$ follows by analyzing the proof of Lemma 6.3.

Suppose, in addition, that $\Psi$ is an orthogonal multiwavelet. Given a sequence $(c_{j,k}^l)_{(l,j,k)\in\Lambda}$ satisfying (6.4) by Lemma 6.3, $\sum_{(l,j,k)\in\Lambda} c_{j,k}^l \psi_{j,k}^l$ converges unconditionally to some element $f \in H_A^p$. Since $\psi^l$'s belong to the dual of $H_A^p$ we conclude that for any $(l', j', k') \in \Lambda$

$$\langle f, \psi_{k',j'}^{l'} \rangle = \left\langle \sum_{(l,j,k)\in\Lambda} c_{j,k}^l \psi_{j,k}^l, \psi_{j',k'}^{l'} \right\rangle = \sum_{(l,j,k)\in\Lambda} c_{j,k}^l \langle \psi_{j,k}^l, \psi_{j',k'}^{l'} \rangle = c_{j',k'}^{l'}.$$

Finally, (6.14) holds by Lemma 6.3. $\square$

**Wavelet expansion coefficients for $L^p$.** It comes as no surprise that Theorems 5.8, 5.10, and 6.7 can be extended for exponents $p > 1$. In fact, orthogonal wavelets with very mild decay conditions already form an unconditional basis for $L^p$, $1 < p < \infty$, see [Po]. Analogously, we can characterize the sequence space of wavelet expansion coefficients for $L^p$. This sequence space is given by exactly the same formula, e.g., (6.4) or (6.5). To show this, we will need two simple lemmas which hold for arbitrary measures, see [Wo2, Corollary 7.10]

LEMMA 6.8. *Suppose $0 < p < \infty$ and $(f_i)_{i\in\mathbb{N}} \subset L^p$. If $\sum_{i\in\mathbb{N}} f_i$ converges unconditionally in $L^p$ then $\|(\sum_{i\in\mathbb{N}} |f_i|^2)^{1/2}\|_p < \infty$. Moreover,*

$$(6.15) \qquad \|(\sum_{i\in\mathbb{N}} |f_i|^2)^{1/2}\|_p \leq (A_p)^{-1/p} \sup_{t\in[0,1]} \left\| \sum_{i\in\mathbb{N}} r_i(t) f_i \right\|_p.$$

PROOF. By Khinchin's inequality and the Fubini Theorem we have

$$A_p \left\| \left( \sum_{i\in F} |f_i|^2 \right)^{1/2} \right\|_p^p \leq \int_0^1 \left\| \sum_{i\in F} r_i(t) f_i \right\|_p^p dt \leq B_p \left\| \left( \sum_{i\in F} |f_i|^2 \right)^{1/2} \right\|_p^p,$$

for any finite $F \subset \mathbb{N}$. If $\sum_{i\in\mathbb{N}} f_i$ converges unconditionally in $L^p$ then by Lemma 5.4,

$$C = \sup_{t\in[0,1]} \left\| \sum_{i\in\mathbb{N}} r_i(t) f_i \right\|_p < \infty.$$

Hence, $A_p \|(\sum_{i\in F} |f_i|^2)^{1/2}\|_p^p \leq C^p$. Since $F \subset \mathbb{N}$ is arbitrary this shows (6.15). $\square$

LEMMA 6.9. *Suppose $0 < p < \infty$ and $(f_i)_{i \in \mathbb{N}} \subset L^p$ is such that $\|f_i\|_p \geq \delta$ for some $\delta > 0$ and all $i \in \mathbb{N}$. Then there is a sequence of coefficients $(\epsilon_i)_{i \in \mathbb{N}}$, $\epsilon_i = \pm 1$ such that*

$$\limsup_{N \to \infty} \left\| \sum_{i=1}^{N} \epsilon_i f_i \right\|_p = \infty. \tag{6.16}$$

PROOF. We claim that it suffices to show

$$\int \left( \sum_{i \in \mathbb{N}} |f_i|^2 \right)^{p/2} = \infty. \tag{6.17}$$

Assume for the time being that (6.17) holds. We claim that for any $j \in \mathbb{N}$ and $r > 0$ there is $N \geq j$ and a sequence $\epsilon_j, \ldots, \epsilon_N$ with $\epsilon_i = \pm 1$ such that

$$\left\| \sum_{i=j}^{N} \epsilon_i f_i \right\|_p > r.$$

Indeed, it suffices to take $N$ such that $\|(\sum_{i=j}^{N} |f_i|^2)\|_p > r(A_p)^{-1/p}$ and apply (6.15). By a simple induction we can now produce a sequence $(\epsilon_i)_{i \in \mathbb{N}}$, $\epsilon_i = \pm 1$ satisfying (6.16).

In order to show (6.17) we need to consider two cases. Suppose first that

$$\limsup_{i \to \infty} |\{x : |f_i(x)| > \eta\}| = 0 \qquad \text{for every } \eta > 0. \tag{6.18}$$

If (6.18) holds then there is a subsequence $(f_{i_j})$ and a decreasing sequence of positive numbers $1 = \eta_0 > \eta_1 > \ldots$, such that

$$\int_{\eta_j < |f_{i_j}| < \eta_{j-1}} |f_{i_j}|^p > \delta/2 \qquad \text{for all } j = 1, 2, \ldots \tag{6.19}$$

Indeed, we are going to proceed by induction. Let $i_1$ be such that $\int_{|f_{i_1}|<1} |f_{i_1}|^p > 3/4\delta$. Choose $\eta_1$ such that $\int_{|f_{i_1}| \leq \eta_1} |f_{i_1}|^p < 1/4\delta$. Hence, (6.19) holds for $j = 1$. Assume (6.19) holds up to some $j > 1$. By (6.18) we can pick $i_{j+1}$ such that $\int_{|f_{i_{j+1}}|<\eta_j} |f_{i_{j+1}}|^p > 3/4\delta$. Choose $\eta_{j+1}$ such that $\int_{|f_{i_{j+1}}| \leq \eta_{j+1}} |f_{i_{j+1}}|^p < 1/4\delta$. This completes the induction step since

$$\int_{\eta_{j+1} < |f_{i_{j+1}}| < \eta_j} |f_{i_{j+1}}|^p > \delta/2.$$

Define sets $S_j = \{x : \eta_{j+1} < |f_{i_{j+1}}(x)| < \eta_j\}$. Again by induction we can find a subsequence $(j_k)$ such that

$$\int_{S_{j_k} \setminus (S_{j_1} \cup \ldots \cup S_{j_{k-1}})} |f_{i_{j_k}}|^p > \delta/4 \qquad \text{for all } k = 1, 2, \ldots \tag{6.20}$$

Indeed, by (6.18) and (6.19) for any set $S$ with $|S| < \infty$,

$$\liminf_{j \to \infty} \int_{S_j \setminus S} |f_{i_j}|^p \geq \delta/2,$$

since $\eta_j \to 0$ as $j \to \infty$. Therefore by (6.20),

$$\int \left(\sum_{i\in\mathbb{N}} |f_i|^2\right)^{p/2} \geq \sum_{k\in\mathbb{N}} \int_{S_{j_k}\setminus(S_{j_1}\cup\ldots S_{j_{k-1}})} |f_{i_{j_k}}|^p = \infty.$$

On the other hand, if (6.18) fails then there exist $\eta, \mu > 0$ such that $|\{x : |f_i(x)| > \eta\}| > \mu$ for infinitely many $i$'s. After taking a subsequence we can assume that this holds for all $i \in \mathbb{N}$. Let $S_i = \{x : |f_i(x)| > \eta\}$. If $|\bigcup_{i\in\mathbb{N}} S_i| < \infty$ then necessarily $\sum_{i\in\mathbb{N}} \mathbf{1}_{S_i}(x) = \infty$ on a set of positive measure. Therefore, the square function in (6.17) is infinite on a set of positive measure. Otherwise, if $|\bigcup_{i\in\mathbb{N}} S_i| = \infty$ then the square function is at least $\eta$ on a set of infinite measure. This shows (6.17) and ends the proof. $\square$

The analog of Lemma 6.3 in the range $1 < p < \infty$ is the following lemma.

LEMMA 6.10. *Suppose $1 < p < \infty$ and $\Psi$ is an 1-regular Bessel multiwavelet consisting of nonzero functions. Given a sequence of scalars $(c_{j,k}^l)_{(l,j,k)\in\Lambda}$, the following are equivalent:*

(6.21) $$\sum_{(l,j,k)\in\Lambda} c_{j,k}^l \psi_{j,k}^l \quad \text{converges unconditionally in } L^p,$$

(6.22) $$\left(\sum_{(l,j,k)\in\Lambda} |c_{j,k}^l|^2 |\psi_{j,k}^l(x)|^2\right)^{1/2} \in L^p(\mathbb{R}^n),$$

(6.23) $$\left(\sum_{(l,j,k)\in\Lambda} |c_{j,k}^l|^2 |(\mathbf{1}_{E_l})_{j,k}(x)|^2\right)^{1/2} \in L^p(\mathbb{R}^n),$$

*for any (or some) bounded measurable sets $E_l \subset \mathbb{R}^n$ with $|E_l| > 0$, $l = 1, \ldots, L$.*

Before we present the proof we need to introduce the standardized version of the sequence space under consideration. We also need a description of its dual which can be found in [NT, Section 9].

DEFINITION 6.11. Given $0 < p < \infty$, a dilation $A$, and $L \in \mathbb{N}$ we define $\ell_A^p = \ell_A^p(\Lambda)$ as the space of all sequences $c = (c_{j,k}^l)_{(l,j,k)\in\Lambda}$ such that

(6.24)
$$\|c\|_{\ell_A^p} = \left\|\left(\sum_{(l,j,k)\in\Lambda} |c_{j,k}^l|^2 |(\mathbf{1}_{[0,1]^n})_{j,k}|^2\right)^{1/2}\right\|_p$$
$$= \left\|\left(\sum_{l=1}^L \sum_{I\in\mathcal{I}} |c_I^l|^2 |I|^{-1} \mathbf{1}_I\right)^{1/2}\right\|_p < \infty.$$

If the dilation $A = 2Id$ and $L = 1$ then the sequence space $\ell_A^p$ coincides with the space $\dot{\mathbf{f}}_p^{0,2}$ introduced by Frazier and Jawerth in [FJ2]. Obviously we could have defined a whole scale of spaces $\dot{\mathbf{f}}_p^{\alpha,q}$ for a general dilation $A$. However, the investigation of anisotropic homogeneous Triebel-Lizorkin spaces $\dot{\mathbf{F}}_p^{\alpha,q}$ is beyond the scope of this work. For the isotropic theory we refer the interested reader to [FJW, Tr1, Tr2]; for the function space theory on spaces of homogeneous type, see [Hn, HS, HW].

# 6. CHARACTERIZATION OF $H^p$ IN TERMS OF WAVELET COEFFICIENTS

LEMMA 6.12. *Suppose $A$ is a dilation and $1 < p < \infty$. The dual of $\ell_A^p$ is isometrically isomorphic to $\ell_A^q$, where $1/p + 1/q = 1$. The duality is given by*

$$(6.25) \qquad \langle c, d \rangle = \sum_{(l,j,k) \in \Lambda} c_{j,k}^l \overline{d_{j,k}^l} = \sum_{l=1}^{L} \sum_{I \in \mathcal{I}} c_I^l \overline{d_I^l} \qquad \text{for } c \in \ell_A^p, d \in \ell_A^q.$$

PROOF. Consider the space $L^p(\ell^2) = L^p(\mathbb{R}^n, \ell^2(\mathbb{Z})^{\oplus L})$ of functions $f : \mathbb{R}^n \to \ell^2(\mathbb{Z})^{\oplus L}$. The norm of $f = (f^1, \ldots, f^L)$, where $f^l : \mathbb{R}^n \to \ell^2(\mathbb{Z})$ in this space is given by

$$\|f\|_{L^p(\ell^2)} = \left( \int_{\mathbb{R}^n} \left( \sum_{l=1}^{L} \|f^l(x)\|_{\ell^2(\mathbb{Z})}^2 \right)^{p/2} dx \right)^{1/p}.$$

We can identify $\ell_A^p$ as a subspace of $L^p(\ell^2)$. Indeed, given $c \in \ell_A^p$ consider the function $f = (f^1, \ldots, f^L)$ whose $j$'th coordinate is given by

$$(6.26) \qquad f^l(x)_j = \sum_{\substack{I \in \mathcal{I} \\ \text{scale}(I) = j}} c_I^l |I|^{-1/2} \mathbf{1}_I(x) \qquad \text{for } l = 1, \ldots, L.$$

This identification defines an isometric inclusion of $\ell_A^p$ into $L^p(\ell^2)$. Functional analysis tells us that the dual of $L^p(\ell^2)$ is isometrically isomorphic to $L^q(\ell^2)$, where $1/p + 1/q = 1$. Furthermore, the duality is given by

$$(6.27) \qquad \langle f, g \rangle = \int_{\mathbb{R}^n} \sum_{l=1}^{L} \langle f^l(x), g^l(x) \rangle dx \qquad \text{for } f \in L^p(\ell^2), g \in L^q(\ell^2).$$

Also $(\ell_A^p)^*$ is isomorphic with the quotient of $L^q(\ell^2)$ and the annihilator of $\ell_A^p$, which consists of functions $f \in L^q(\ell^2)$ with $\int_I f^l(x)_j dx = 0$ for all $l = 1, \ldots, L$, $I \in \mathcal{I}$, $j = \text{scale}(I)$. Therefore, the dual of $\ell_A^p$ can be identified with $\ell_A^q$. Moreover, the dual of the inclusion $\ell_A^p \hookrightarrow L^p(\ell^2)$ is a projection $P : L^q(\ell^2) \to (\ell_A^p)^*$ given by

$$(Pf)(x)_j = \sum_{l=1}^{L} \sum_{\substack{I \in \mathcal{I} \\ \text{scale}(I) = j}} \left( |I|^{-1} \int_I f^l(y)_i dy \right) \mathbf{1}_I.$$

The duality (6.25) follows from (6.26) and (6.27). $\square$

PROOF OF LEMMA 6.10. (6.21) $\implies$ (6.22) follows from Lemma 6.8. (6.22) $\iff$ (6.23) is shown in the same way as in Lemma 6.3. However, (6.23) $\implies$ (6.21) is shown in a different manner since we do not have an atomic decomposition at our disposal for $p > 1$.

By Lemma 5.5 for any sequence of bounded scalars $\epsilon = (\epsilon_{j,k}^l)_{(l,j,k) \in \Lambda}$ the operator $T_\epsilon : L^2(\mathbb{R}^n) \to L^2(\mathbb{R}^n)$ given by

$$(6.28) \qquad T_\epsilon(f) = \sum_{(l,j,k) \in \Lambda} \epsilon_{j,k}^l \langle f, \psi_{j,k}^l \rangle \psi_{j,k}^l,$$

is Calderón-Zygmund of order 1. Hence, $T_\epsilon$ is a Calderón-Zygmund operator in the sense of Definition 9.1, Chapter 1. Therefore, $T_\epsilon$ extends to a bounded operator on $L^p$ for $1 < p < \infty$. Furthermore, if $|\epsilon_{j,k}^l| \leq 1$ then $T_\epsilon$'s are uniformly bounded on $L^p$. We are going to show that (6.28) holds also for $f \in L^p$ with unconditional convergence in $L^p$.

If the series $\sum_{(l,j,k)\in\Lambda}\langle f,\psi^l_{j,k}\rangle\psi^l_{j,k}$ does not converge unconditionally in $L^p$ then there are pairwise disjoint finite sets $F_i \subset \Lambda$ and coefficients $\epsilon^l_{j,k} = \pm 1$, $(l,j,k) \in F_i$ such that
$$f_i = \sum_{(l,j,k)\in F_i} \epsilon^l_{j,k} \langle f, \psi^l_{j,k}\rangle \psi^l_{j,k},$$
satisfies $\|f_i\|_p > \delta$ for some $\delta > 0$ and all $i \in \mathbb{N}$. By Lemma 6.9, there are coefficients $\epsilon_i = \pm 1$ such that $\limsup_{N\to\infty} \|\sum_{i=1}^N \epsilon_i f_i\|_p = \infty$. This clearly contradicts the uniform boundedness of $T_\epsilon$'s. Finally, $\sum_{(l,j,k)\in\Lambda} \epsilon^l_{j,k}\langle f,\psi^l_{j,k}\rangle \psi^l_{j,k}$ converges to $T_\epsilon f$ by considering $f \in L^2 \cap L^p$. By Lemma 6.8,

$$\int_{\mathbb{R}^n} \left( \sum_{(l,j,k)\in\Lambda} |\langle f,\psi^l_{j,k}\rangle|^2 |\psi^l_{j,k}(x)|^2 \right)^{p/2} dx$$

$$\leq (A_p)^{-1} \sup_{t\in[0,1]} \left\| \sum_{(l,j,k)\in\Lambda} \epsilon^l_{j,k}(t)\langle f,\psi^l_{j,k}\rangle \psi^l_{j,k} \right\|_p^p \leq C/A_p \|f\|_p^p,$$

for any $f \in L^p$. Therefore, $\|(\langle f,\psi^l_{j,k}\rangle)\|_{\ell^p_A}^p \leq C\|f\|_p$. In other words, the analysis operator is bounded from $L^p$ to $\ell^p_A$. By Lemma 6.12, the dual of analysis operator is a synthesis operator

$$\ell^q_A \ni (c^l_{j,k})_{(l,j,k)\in\Lambda} \mapsto \sum_{(l,j,k)\in\Lambda} c^l_{j,k} \psi^l_{j,k} \in L^q(\mathbb{R}^n),$$

where $1/p + 1/q = 1$. We also have

$$\left\| \sum_{(l,j,k)\in\Lambda} c^l_{j,k} \psi^l_{j,k} \right\|_q \leq C\|c\|_{\ell^q_A} \qquad \text{for all } c \in \ell^q_A.$$

This shows (6.23) $\implies$ (6.21). $\qquad\square$

We are now ready to prove the characterization of the sequence space of wavelet expansion coefficients of functions in $L^p$.

THEOREM 6.13. *Suppose $1 < p < \infty$, and $\Psi$ is a 1-regular tight frame multi-wavelet. Then the series $\sum_{(l,j,k)\in\Lambda}\langle f,\psi^l_{j,k}\rangle\psi^l_{j,k}$ converges unconditionally to $f$ in $L^p$ for any $f \in L^p$, and*

(6.29)
$$\|f\|_{L^p} \sim \left\|\left(\sum_{(l,j,k)\in\Lambda} |\langle f,\psi^l_{j,k}\rangle|^2 |\psi^l_{j,k}(\cdot)|^2 \right)^{1/2}\right\|_{L^p}$$
$$\sim \left\|\left(\sum_{(l,j,k)\in\Lambda} |\langle f,\psi^l_{j,k}\rangle|^2 |(\mathbf{1}_{E_l})_{j,k}(\cdot)|^2 \right)^{1/2}\right\|_{L^p},$$

*for any (or some) bounded measurable sets $E_l \subset \mathbb{R}^n$ with $|E_l| > 0$, $l = 1, \ldots, L$. The equivalence constants in (6.29) do not depend on $f$.*

*If, in addition, $\Psi$ is an orthonormal multiwavelet then for any $(c^l_{j,k})_{(l,j,k)\in\Lambda}$ satisfying*

(6.4)
$$\left( \sum_{(l,j,k)\in\Lambda} |c^l_{j,k}|^2 |\psi^l_{j,k}(x)|^2 \right)^{1/2} \in L^p(\mathbb{R}^n),$$

there is a unique $f \in L^p$ such that $c_{j,k}^l = \langle f, \psi_{j,k}^l \rangle$ for all $(l, j, k) \in \Lambda$ and (6.29) holds.

PROOF. The first part of the theorem follows from Lemma 6.10, see the proof of (6.23) $\implies$ (6.21).

Suppose, in addition, that $\Psi$ is an orthonormal multiwavelet. If $(c_{j,k}^l)_{(l,j,k)\in\Lambda}$ satisfies (6.4) then by Lemma 6.10, $\sum_{(l,j,k)\in\Lambda} c_{j,k}^l \psi_{j,k}^l$ converges unconditionally to some $f \in L^p$. Since $\psi^l$'s belong to the dual of $L^p$ we must have $c_{j,k}^l = \langle f, \psi_{j,k}^l \rangle$ for all $(l, j, k) \in \Lambda$. This completes the proof of Theorem 6.13. $\square$

# Notation Index

| | |
|---|---|
| $\|\cdot\|$ | the standard norm in $\mathbb{R}^n$ or the Lebesgue measure of a subset in $\mathbb{R}^n$ |
| $\mathbf{B}(z,r)$ | the Euclidean ball with center $z$ and radius $r$ |
| $\langle \cdot, \cdot \rangle$ | the scalar product in $\mathbb{R}^n$, $L^2(\mathbb{R}^n)$, or $\langle f, \varphi \rangle = f(\varphi)$ for $f \in \mathcal{S}'$, $\varphi \in \mathcal{S}$ |
| $\{e_i\}_{i=1}^n$ | the standrad orthonormal basis in $\mathbb{R}^n$ |
| $\hat{f}(\xi)$ | the Fourier transform of $f$, $\hat{f}(\xi) = \int_{\mathbb{R}^n} f(x) e^{-2\pi i \langle x, \xi \rangle} dx$ |
| supp $f$ | the support of $f$, supp $f = \{x : f(x) \neq 0\}$ |
| $\mathfrak{D}^k f(x)$ | the derivative of $f$ at the point $x$ thought of as a symmetric multilinear operator $(\mathbb{R}^n)^k \to \mathbb{R}^n$ |
| $\mathbf{1}_E$ | the indicator function of the set $E \subset \mathbb{R}^n$ |
| $A$ | the dilation, i.e., $n \times n$ matrix with all eigenvalues $\lambda$, $|\lambda| > 1$ |
| $\lambda_-$ | the absolute value of the smallest eigenvalue of $A$ |
| $\lambda_+$ | the absolute value of the biggest eigenvalue of $A$ |
| $b$ | the number equal to $|\det A|$ |
| $B_k$ | the dilated balls of the form $B_k = A^k \Delta$, $k \in \mathbb{Z}$, where $\Delta$ is a special ellipsoid tailored to the dilation $A$ with $|\Delta| = 1$ |
| $\mathcal{B}$ | the family of dilated balls of the form $x_0 + B_k$, $x_0 \in \mathbb{R}^n$, $k \in \mathbb{Z}$ |
| $\omega$ | the smallest integer such that $2B_0 \subset B_\omega$ |
| $\rho$ | the quasi-norm associated to the dilation $A$ |
| $\varphi_k$ | the dilate of $\varphi$ to the scale $k \in \mathbb{Z}$, $\varphi_k(x) = b^{-k} \varphi(A^{-k} x)$ |
| $\psi_{j,k}$ | the dilate and translate of $\psi$ given by $\psi_{j,k}(x) = |\det A|^{j/2} \psi(A^j x - k)$ where $j \in \mathbb{Z}$, $k \in \mathbb{Z}^n$ |
| $\Psi$ | the collection $\Psi = \{\psi^1, \ldots, \psi^L\} \subset L^2(\mathbb{R}^n)$ |
| $\mathcal{I}$ | the collection of dilated cubes $\{A^j([0,1]^n + k) : j \in \mathbb{Z}, k \in \mathbb{Z}^n\}$ |
| $\psi_I$ | is $\psi_{-j,k}$ for $I = A^j([0,1]^n + k) \in \mathcal{I}$; the exception to this rule is $\mathbf{1}_I$ |
| $\Lambda$ | the index set $\Lambda = \{(l,j,k) : l = 1, \ldots, L, j \in \mathbb{Z}, k \in \mathbb{Z}^n\}$ |
| $D_A$ | the dilation operator (usually on $H^p$), $D_A f(x) = |\det A|^{1/p} f(Ax)$ |
| $\tau_y$ | the translation operator, $\tau_y f(x) = f(x - y)$ |
| $(p, q, s)$ | an admissible triplet with respect to the dilation $A$, i.e., $0 < p \leq 1$, $1 \leq q \leq \infty$, $p < q$, $s \in \mathbb{N}$, and $s \geq \lfloor (1/p - 1) \ln b / \ln \lambda_- \rfloor$ |
| $M_\varphi f$ | the nontangential maximal function of $f$ with respect to $\varphi$ |
| $M_\varphi^0 f$ | the radial maximal function of $f$ with respect to $\varphi$ |
| $T_\varphi^N f(x)$ | the tangential maximal function of $f$ with respect to $\varphi$ |
| $M_N f$ | the nontangential grand maximal function of $f$ |
| $M_N^0 f$ | the radial grand maximal function of $f$ |
| $M_{HL} f$ | the Hardy-Littlewood maximal function of $f$ |
| $C^r$ | the space of functions with continuous partial derivatives up to order $r$, $r = 0, 1, \ldots, \infty$ |

# NOTATION INDEX

| | |
|---|---|
| $\mathcal{S}$ | the space of test functions (the Schwartz class) |
| $\mathcal{S}'$ | the space of tempered distributions |
| $\mathcal{S}_N$ | the subset of $\mathcal{S}$ consisting of all $\varphi$ satisfying $\sup_{x\in\mathbb{R}^n}\sup_{|\alpha|\le N}\max(1,\rho(x)^N)|\partial^\alpha\varphi(x)|\le 1$ |
| $\mathcal{P}_s$ | the space of polynomials of degree $\le s$ |
| $L^p$ | the space of functions with $\int_{\mathbb{R}^n}|f(x)|^p dx<\infty$, $0<p\le\infty$ |
| $\|f\|_p$ | the quasi-norm in $L^p$, $\|f\|_p=(\int_{\mathbb{R}^n}|f(x)|^p dx)^{1/p}$ |
| $H^p$ | the anisotropic Hardy space $H^p_A$ associated with the dilation $A$ |
| $\|f\|_{H^p}$ | the norm of $f\in H^p$, $\|f\|_{H^p}=\|M_N f\|_p$ |
| $H^p_{q,s}$ | the atomic anisotropic Hardy space associated with the dilation $A$ for an admissible triplet $(p,q,s)$ |
| $C^l_{q,s}$ | the Campanato space for $l\ge 0$, $1\le q\le\infty$, $s=0,1,\ldots$ |
| $\Theta^q_s$ | the space of functions in $L^q$ with bounded support and vanishing moments up to order $s$ |

# Bibliography

[Au1]   P. Auscher, *Wavelet bases for $L^2(\mathbb{R})$ with rational dilation factor*, Wavelets and their applications, Jones and Bartlett, Boston, MA, 1992, pp. 439–451.

[Au2]   P. Auscher, *Toute base d'ondelettes régulières de $L^2(\mathbb{R})$ est issue d'une analyse multi-résolution régulière*, C. R. Acad. Sci. Paris Sér. I Math. **315** (1992), 1227–1230.

[Au3]   P. Auscher, *Solution of two problems on wavelets*, J. Geom. Anal. **5** (1995), 181–236.

[Ay1]   A. Ayache, *Construction of non separable dyadic compactly supported orthonormal wavelet bases for $L^2(\mathbb{R}^2)$ of arbitrarily high regularity*, Rev. Mat. Iberoamericana **15** (1999), 37–58.

[Ay2]   A. Ayache, *Some methods for constructing nonseparable, orthonormal, compactly supported wavelet bases*, Appl. Comput. Harmon. Anal. **10** (2001), 99–111.

[BS]    A. Baernstein and E. T. Sawyer, *Embedding and multiplier theorems for $H^p(\mathbb{R}^n)$*, Mem. Amer. Math. Soc. **53** (1985), no. 318.

[Ba]    L. W. Baggett, *An abstract interpretation of wavelet dimension function using group representations*, J. Funct. Anal. **173** (2000), 1–20.

[Bt]    G. Battle, *Phase space localization theorem for ondelettes*, J. Math. Phys. **30** (1989), 2195–2196.

[BW]    E. Belogay and Y. Wang, *Arbitrarily smooth orthogonal nonseparable wavelets in $\mathbb{R}^2$*, SIAM J. Math. Anal. **30** (1999), 678–697.

[BDR]   C. de Boor, R. A. DeVore, and A. Ron, *The structure of finitely generated shift-invariant spaces in $L_2(\mathbb{R}^d)$*, J. Funct. Anal. **119** (1994), 37–78.

[Bo1]   M. Bownik, *Tight frames of multidimensional wavelets*, J. Fourier Anal. Appl. **3** (1997), 525–542.

[Bo2]   M. Bownik, *A characterization of affine dual frames in $L^2(\mathbb{R}^n)$*, Appl. Comp. Harm. Anal. **8** (2000), 203–221.

[Bo3]   M. Bownik, *The structure of shift invariant subspaces of $L^2(\mathbb{R}^n)$*, J. Funct. Anal. **177** (2000), 282–309.

[Bo4]   M. Bownik, *The construction of r-regular wavelets for arbitrary dilations*, J. Fourier Anal. Appl. **7** (2001), 489–506.

[Bo5]   M. Bownik, *Combined MSF multiwavelets*, J. Fourier Anal. Appl. **8** (2002), 201–210.

[Bo6]   M. Bownik, *On a problem of Daubechies*, Constr. Approx. (to appear).

[BRS]   M. Bownik, Z. Rzeszotnik, and D. Speegle, *A characterization of dimension functions of wavelets*, Appl. Comput. Harmon. Anal. **10** (2001), 71–92.

[BR]    M. Bownik and Z. Rzeszotnik, *The spectral function of shift-invariant spaces*, preprint (2002).

[BS1]   M. Bownik and D. Speegle, *Meyer type wavelet bases in $\mathbb{R}^2$*, J. Approx. Theory **116** (2002), 49–75.

[BS2]   M. Bownik and D. Speegle, *The wavelet dimension function for real dilations and dilations admitting non-MSF wavelets*, Approximation Theory X: Wavelets, Splines, and Applications, Vanderbilt University Press, 2002, pp. 63-85.

[BGS]   D. L. Burkholder, R. F. Gundy, and M. L. Silverstein, *A maximal function characterization of the class $H^p$*, Trans. Amer. Math. Soc. **157** (1971), 137–153.

[Ca]    A.-P. Calderón, *An atomic decomposition of distributions in parabolic $H^p$ spaces*, Advances in Math. **25** (1977), 216–225.

[CT1]   A.-P. Calderón and A. Torchinsky, *Parabolic maximal functions associated with a distribution*, Advances in Math. **16** (1975), 1–64.

[CT2]   A.-P. Calderón and A. Torchinsky, *Parabolic maximal functions associated with a distribution. II*, Advances in Math. **24** (1977), 101–171.

[Cm]   S. Campanato, *Proprietà di una famiglia di spazi funzionali*, Ann. Scuola Norm. Sup. Pisa (3) **18** (1964), 137–160.

[Cz]   P. G. Casazza, *The art of frame theory*, Taiwanese J. Math. **4** (2000), 129–201.

[Ch]   M. Christ, *Lectures on singular integral operators*, Published for the Conference Board of the Mathematical Sciences, Washington, DC, 1990.

[CS]   C. K. Chui and X. Shi, *Orthonormal wavelets and tight frames with arbitrary real dilations*, Appl. Comput. Harmon. Anal. **9** (2000), 243–264.

[CMW]  C. K. Chui, W. Czaja, M. Maggioni, and G. Weiss, *Characterization of general tight wavelet frames with matrix dilations and tightness preserving oversampling*, J. Fourier Anal. Appl. **8** (2002), 173–200.

[CGV]  A. Cohen, K. Gröchenig, and L. F. Villemoes, *Regularity of multivariate refinable functions*, Constr. Approx. **15** (1999), 241–255.

[Co]   R. R. Coifman, *A real variable characterization of $H^p$*, Studia Math. **51** (1974), 269–274.

[CMS]  R. R. Coifman, Y. Meyer, and E. M. Stein, *Some new function spaces and their applications to harmonic analysis*, J. Funct. Anal. **62** (1985), 304–335.

[CR]   R. R. Coifman and R. Rochberg, *Representation theorems for holomorphic and harmonic functions in $L^p$*, Representation theorems for Hardy spaces, Soc. Math. France, Paris, 1980, pp. 11–66.

[CRW]  R. R. Coifman, R. Rochberg, and G. Weiss, *Factorization theorems for Hardy spaces in several variables*, Ann. of Math. **103** (1976), 611–635.

[CW1]  R. R. Coifman and G. Weiss, *Analyse harmonique non-commutative sur certains espaces homogènes*, Étude de certaines intégrales singulières, Lecture Notes in Mathematics, Vol. 242, Springer-Verlag, Berlin, 1971.

[CW2]  R. R. Coifman and G. Weiss, *Extensions of Hardy spaces and their use in analysis*, Bull. Amer. Math. Soc. **83** (1977), 569–645.

[DL]   X. Dai and D. R. Larson, *Wandering vectors for unitary systems and orthogonal wavelets*, Mem. Amer. Math. Soc. **134** (1998), no. 640.

[DLS]  X. Dai, D. R. Larson, and D. M. Speegle, *Wavelet sets in $\mathbb{R}^n$*, J. Fourier Anal. Appl. **3** (1997), 451–456.

[Da1]  I. Daubechies, *Orthonormal bases of compactly supported wavelets*, Comm. Pure Appl. Math. **41** (1988), 909–996.

[Da2]  I. Daubechies, *Ten lectures on wavelets*, Society for Industrial and Applied Mathematics (SIAM), Philadelphia, PA, 1992.

[DJ]   G. David and J.-L. Journé, *A boundedness criterion for generalized Calderón-Zygmund operators*, Ann. of Math. **120** (1984), 371–397.

[DJS]  G. David, J.L. Journé, and S. Semmes, *Opérateurs de Calderón-Zygmund, fonctions para-accrétives et interpolation*, Rev. Mat. Iberoamericana **1** (1985), 1–56.

[Dv]   G. David, *Wavelets and singular integrals on curves and surfaces*, Springer-Verlag, Berlin, 1991.

[DS]   R. J. Duffin and A. C. Schaeffer, *A class of nonharmonic Fourier series*, Trans. Amer. Math. Soc. **72** (1952), 341–366.

[DRS]  P. L. Duren, B. W. Romberg, and A. L. Shields, *Linear functionals on $H^p$ spaces with $0 < p < 1$*, J. Reine Angew. Math. **238** (1969), 32–60.

[Du]   P. L. Duren, *Theory of $H^p$ spaces*, Pure and Applied Mathematics, Vol. 38, Academic Press, New York, 1970.

[Fe]   C. Fefferman, *Harmonic analysis and $H^p$ spaces*, Studies in harmonic analysis (Proc. Conf., DePaul Univ., Chicago, Ill., 1974), Math. Assoc. Amer., Washington, D.C., 1976, pp. 38–75. MAA Stud. Math., Vol. 13.

[FS1]  C. Fefferman and E. M. Stein, *Some maximal inequalities*, Amer. J. Math. **93** (1971), 107–115.

[FS2]  C. Fefferman and E. M. Stein, *$H^p$ spaces of several variables*, Acta Math. **129** (1972), 137–193.

[FoS]  G. B. Folland and E. M. Stein, *Hardy spaces on homogeneous groups*, Princeton University Press, Princeton, N.J., 1982.

[FJ1]  M. Frazier and B. Jawerth, *A discrete transform and decompositions of distribution spaces*, J. Funct. Anal. **93** (1990), 34–170.

# BIBLIOGRAPHY

[FJ2] M. Frazier and B. Jawerth, *Applications of the $\phi$ and wavelet transforms to the theory of function spaces*, Wavelets and their applications, Jones and Bartlett, Boston, MA, 1992, pp. 377–417.

[FJW] M. Frazier, B. Jawerth, and G. Weiss, *Littlewood-Paley theory and the study of function spaces*, Published for the Conference Board of the Mathematical Sciences, Washington, DC, 1991.

[GR] J. García-Cuerva and J. L. Rubio de Francia, *Weighted norm inequalities and related topics*, North-Holland Mathematics Studies, vol. 116, North-Holland Publishing Co., Amsterdam, 1985.

[Ga] J. B. Garnett, *Bounded analytic functions*, Academic Press Inc., New York, 1981.

[GL] J. B. Garnett and R. H. Latter, *The atomic decomposition for Hardy spaces in several complex variables*, Duke Math. J. **45** (1978), 815–845.

[Gt] A. E. Gatto, *An atomic decomposition of distributions in parabolic $H^p$ spaces*, Rev. Un. Mat. Argentina **29** (1979/80), 169–179.

[Ge] D. Geller, *Some results in $H^p$ theory for the Heisenberg group*, Duke Math. J. **47** (1980), 365–390.

[Gr] H. Greenwald, *On the theory of homogeneous Lipschitz spaces and Campanato spaces*, Pacific J. Math. **106** (1983), 87–93.

[GM] K. Gröchenig and W. R. Madych, *Multiresolution analysis, Haar bases, and self-similar tilings of $\mathbb{R}^n$*, IEEE Trans. Inform. Theory **38** (1992), 556–568.

[Ha] A. Haar, *Zur Theorie der orthogonalen Funktionensysteme*, Math. Ann. **69** (1910), 331–371.

[Hn] Y. S. Han, *Triebel-Lizorkin spaces on spaces of homogeneous type*, Studia Math. **108** (1994), 247–273.

[HS] Y. S. Han and E. T. Sawyer, *Littlewood-Paley theory on spaces of homogeneous type and the classical function spaces*, Mem. Amer. Math. Soc. **110** (1994), no. 530.

[HW] Y. S. Han and G. Weiss, *Function spaces on spaces of homogeneous type*, Essays on Fourier analysis in honor of Elias M. Stein (Princeton, NJ, 1991), Princeton Univ. Press, Princeton, NJ, 1995, pp. 211–224.

[HL] D. Han and D. R. Larson, *Frames, bases and group representations*, Mem. Amer. Math. Soc. **147** (2000), no. 697.

[HeW] E. Hernández and G. Weiss, *A first course on wavelets*, Studies in Advanced Mathematics, CRC Press, Boca Raton, FL, 1996.

[Hö] L. Hörmander, *Estimates for translation invariant operators in $L^p$ spaces*, Acta Math. **104** (1960), 93–140.

[JMR] S. Jaffard, Y. Meyer, and R. D. Ryan, *Wavelets. Tools for science & technology*, SIAM, Philadelphia, PA, 2001.

[Ja] S. Janson, *Generalizations of Lipschitz spaces and an application to Hardy spaces and bounded mean oscillation*, Duke Math. J. **47** (1980), 959–982.

[JTW] S. Janson, M. Taibleson, and G. Weiss, *Elementary characterizations of the Morrey-Campanato spaces*, Harmonic analysis (Cortona, 1982), Springer, Berlin, 1983, pp. 101–114.

[JN] F. John and L. Nirenberg, L., *On functions of bounded mean oscillation*, Comm. Pure Appl. Math. **14** (1961), 415–426.

[JW] A. Jonsson and H. Wallin, *Function spaces on subsets of $\mathbb{R}^n$*, Math. Rep. **2** (1984), no. 1.

[KL] J.-P. Kahane and P.-G. Lemarié-Rieusset, *Fourier Series and Wavelets*, Gordon and Breach Publishers, Luxembourg, 1995.

[KLW] N. J. Kalton, C. Leranoz, and P. Wojtaszczyk, *Uniqueness of unconditional bases in quasi-Banach spaces with applications to Hardy spaces*, Israel J. Math. **72** (1990), 299–311.

[KPR] N. J. Kalton, N. T. Peck, and J. W. Roberts, *An F-space sampler*, Cambridge University Press, Cambridge, 1984.

[Ko] P. Koosis, *Introduction to $H_p$ spaces*, Cambridge University Press, Cambridge, 1998.

[KV] A. Korányi and S. Vági, *Singular integrals on homogeneous spaces and some problems of classical analysis*, Ann. Scuola Norm. Sup. Pisa (3) **25** (1971), 575–648 (1972).

[Kr] S. G. Krantz, *Lipschitz spaces, smoothness of functions, and approximation theory*, Exposition. Math. **1** (1983), 193–260.

# BIBLIOGRAPHY

[KN]   L. Kuipers and H. Niederreiter, *Uniform distribution of sequences*, Pure and Applied Mathematics, Wiley-Interscience [John Wiley & Sons], New York, 1974.

[LW1]  J. C. Lagarias and Y. Wang, *Haar type orthonormal wavelet bases in $\mathbb{R}^2$*, J. Fourier Anal. Appl. **2** (1995), 1–14.

[LW2]  J. C. Lagarias and Y. Wang, *Haar Bases for $L^2(\mathbb{R}^n)$ and Algebraic Number Theory*, J. Number Theory **57** (1996), 181–197.

[LW3]  J. C. Lagarias and Y. Wang, *Corrigendum and Addendum to: Haar Bases for $L^2(\mathbb{R}^n)$ and Algebraic Number Theory*, J. Number Theory **76** (1999), 330–336.

[La]   R. H. Latter, *A characterization of $H^p(\mathbb{R}^n)$ in terms of atoms*, Studia Math. **62** (1978), 93–101.

[LU]   R. H. Latter and A. Uchiyama, *The atomic decomposition for parabolic $H^p$ spaces*, Trans. Amer. Math. Soc. **253** (1979), 391–398.

[LR]   P.-G. Lemarié-Rieusset, *Projecteurs invariants, matrices de dilatation, ondelettes et analyses multirésolutions*, Rev. Mat. Iberoamericana **10** (1994), 283–347.

[MS1]  R. A. Macías and C. Segovia, *Lipschitz functions on spaces of homogeneous type*, Adv. in Math. **33** (1979), 257–270.

[MS2]  R. A. Macías and C. Segovia, *A decomposition into atoms of distributions on spaces of homogeneous type*, Adv. in Math. **33** (1979), 271–309.

[MS3]  R. A. Macías and C. Segovia, *Singular integrals on generalized Lipschitz and Hardy spaces*, Studia Math. **65** (1979), 55–75.

[Md1]  W. R. Madych, *Some elementary properties of multiresolution analyses of $L^2(\mathbb{R}^n)$*, Wavelets, Academic Press, Boston, MA, 1992, pp. 259–294.

[Md2]  W. R. Madych, *Orthogonal wavelet bases for $L^2(\mathbb{R}^n)$*, Fourier analysis (Orono, ME, 1992), Dekker, New York, 1994, pp. 243–302.

[Me]   Y. Meyer, *Wavelets and operators*, Cambridge University Press, Cambridge, 1992.

[MC]   Y. Meyer and R. R. Coifman, *Wavelets. Calderón-Zygmund and multilinear operators*, Cambridge University Press, Cambridge, 1997.

[MTW]  Y. Meyer, M. H. Taibleson, and G. Weiss, *Some functional analytic properties of the spaces $B_q$ generated by blocks*, Indiana Univ. Math. J. **34** (1985), 493–515.

[NT]   F. Nazarov and T. Sergei, *The hunt for a Bellman function: applications to estimates for singular integral operators and to other classical problems of harmonic analysis*, Algebra i Analiz **8** (1996), 32–162.

[Po]   W. Pompe, *Unconditional biorthogonal wavelet bases in $L^p(\mathbb{R}^d)$*, Colloq. Math. **92** (2002), 19–34.

[Ro]   S. Rolewicz, *Metric linear spaces*, D. Reidel Publishing Co., Dordrecht, 1985.

[Ru1]  W. Rudin, *Function theory in polydiscs*, W. A. Benjamin, Inc., New York-Amsterdam, 1969.

[Ru2]  W. Rudin, *Function theory in the unit ball of $\mathbb{C}^n$*, Springer-Verlag, New York, 1980.

[Ru3]  W. Rudin, *Functional analysis*, McGraw-Hill Inc., New York, 1991.

[Rz]   Z. Rzeszotnik, *Characterization theorems in the theory of wavelets*, Ph. D. Thesis, Washington University (2000).

[St1]  E. M. Stein, *Singular integrals and differentiability properties of functions*, Princeton Mathematical Series, No. 30, Princeton University Press, Princeton, N.J., 1970.

[St2]  E. M. Stein, *Harmonic analysis: real-variable methods, orthogonality, and oscillatory integrals*, Princeton Mathematical Series, No. 43, Princeton University Press, Princeton, NJ, 1993.

[SW1]  E. M. Stein and G. Weiss, *On the theory of harmonic functions of several variables. I. The theory of $H^p$-spaces*, Acta Math. **103** (1960), 25–62.

[SW2]  E. M. Stein and G. Weiss, *Introduction to Fourier analysis on Euclidean spaces*, Princeton Mathematical Series, No. 32, Princeton University Press, Princeton, N.J., 1971.

[Sr]   R. S. Strichartz, *Wavelets and self-affine tilings*, Constr. Approx. **9** (1993), 327–346.

[Sö]   J. O. Strömberg, *A modified Franklin system and higher-order spline systems on $\mathbb{R}^n$ as unconditional bases for Hardy spaces*, Conference on harmonic analysis in honor of Antoni Zygmund, Vol. I, II (Chicago, Ill., 1981), Wadsworth, Belmont, Calif., 1983, pp. 475–494.

[ST1]  J. O. Strömberg and A. Torchinsky, *Weights, sharp maximal functions and Hardy spaces*, Bull. Amer. Math. Soc. **3** (1980), 1053–1056.

[ST2]  J. O. Strömberg and A. Torchinsky, *Weighted Hardy spaces*, Springer-Verlag, Berlin, 1989.
[Sz]  W. Szlenk, *An introduction to the theory of smooth dynamical systems*, Translated from the Polish by Marcin E. Kuczma, PWN—Polish Scientific Publishers, Warsaw, 1984.
[TW]  M. H. Taibleson and G. Weiss, *The molecular characterization of certain Hardy spaces*, Representation theorems for Hardy spaces, Soc. Math. France, Paris, 1980, pp. 67–149.
[To]  A. Torchinsky, *Real-variable methods in harmonic analysis*, Academic Press Inc., Orlando, Fla., 1986.
[Tr1]  H. Triebel, *Theory of function spaces*, Birkhäuser Verlag, Basel, 1983.
[Tr2]  H. Triebel, *Theory of function spaces. II*, Birkhäuser Verlag, Basel, 1992.
[Uc1]  A. Uchiyama, *A maximal function characterization of $H^p$ on the space of homogeneous type*, Trans. Amer. Math. Soc. **262** (1980), 579–592.
[Uc2]  A. Uchiyama, *On the radial maximal function of distributions*, Pacific J. Math. **121** (1986), 467–483.
[Uc3]  A. Uchiyama, *Hardy spaces on the Euclidean space*, Springer-Verlag, Tokyo, 2001.
[Wa]  T. Walsh, *The dual of $H^p(\mathbb{R}_+^{n+1})$ for $p<1$*, Canad. J. Math. **25** (1973), 567–577.
[Wo1]  P. Wojtaszczyk, *Banach spaces for analysts*, Cambridge University Press, Cambridge, 1991.
[Wo2]  P. Wojtaszczyk, *A mathematical introduction to wavelets*, Cambridge University Press, Cambridge, 1997.
[Wo3]  P. Wojtaszczyk, *Uniqueness of unconditional bases in quasi-Banach spaces with applications to Hardy spaces. II*, Israel J. Math. **97** (1997), 253–280.
[Wo4]  P. Wojtaszczyk, *Wavelets as unconditional bases in $L_p(\mathbb{R})$*, J. Fourier Anal. Appl. **5** (1999), 73–85.
[Yo]  R. M. Young, *An introduction to nonharmonic Fourier series*, Academic Press Inc., San Diego, CA, 2001.
[Zy]  A. Zygmund, A., *Trigonometric series. Vol. I, II*, Cambridge University Press, Cambridge, 1988.

## Editorial Information

To be published in the *Memoirs*, a paper must be correct, new, nontrivial, and significant. Further, it must be well written and of interest to a substantial number of mathematicians. Piecemeal results, such as an inconclusive step toward an unproved major theorem or a minor variation on a known result, are in general not acceptable for publication. Papers appearing in *Memoirs* are generally longer than those appearing in *Transactions*, which shares the same editorial committee.

As of April 1, 2003, the backlog for this journal was approximately 4 volumes. This estimate is the result of dividing the number of manuscripts for this journal in the Providence office that have not yet gone to the printer on the above date by the average number of monographs per volume over the previous twelve months, reduced by the number of volumes published in four months (the time necessary for preparing a volume for the printer). (There are 6 volumes per year, each containing at least 4 numbers.)

A Consent to Publish and Copyright Agreement is required before a paper will be published in the *Memoirs*. After a paper is accepted for publication, the Providence office will send a Consent to Publish and Copyright Agreement to all authors of the paper. By submitting a paper to the *Memoirs*, authors certify that the results have not been submitted to nor are they under consideration for publication by another journal, conference proceedings, or similar publication.

## Information for Authors

*Memoirs* are printed from camera copy fully prepared by the author. This means that the finished book will look exactly like the copy submitted.

The paper must contain a *descriptive title* and an *abstract* that summarizes the article in language suitable for workers in the general field (algebra, analysis, etc.). The *descriptive title* should be short, but informative; useless or vague phrases such as "some remarks about" or "concerning" should be avoided. The *abstract* should be at least one complete sentence, and at most 300 words. Included with the footnotes to the paper should be the 2000 *Mathematics Subject Classification* representing the primary and secondary subjects of the article. The classifications are accessible from www.ams.org/msc/. The list of classifications is also available in print starting with the 1999 annual index of *Mathematical Reviews*. The Mathematics Subject Classification footnote may be followed by a list of *key words and phrases* describing the subject matter of the article and taken from it. Journal abbreviations used in bibliographies are listed in the latest *Mathematical Reviews* annual index. The series abbreviations are also accessible from www.ams.org/publications/. To help in preparing and verifying references, the AMS offers MR Lookup, a Reference Tool for Linking, at www.ams.org/mrlookup/. When the manuscript is submitted, authors should supply the editor with electronic addresses if available. These will be printed after the postal address at the end of the article.

**Electronically prepared manuscripts.** The AMS encourages electronically prepared manuscripts, with a strong preference for $\mathcal{AMS}$-LaTeX. To this end, the Society has prepared $\mathcal{AMS}$-LaTeX author packages for each AMS publication. Author packages include instructions for preparing electronic manuscripts, the *AMS Author Handbook*, samples, and a style file that generates the particular design specifications of that publication series. Though $\mathcal{AMS}$-LaTeX is the highly preferred format of TeX, author packages are also available in $\mathcal{AMS}$-TeX.

Authors may retrieve an author package from e-MATH starting from www.ams.org/tex/ or via FTP to ftp.ams.org (login as anonymous, enter username as password, and type cd pub/author-info). The *AMS Author Handbook* and the *Instruction Manual* are available in PDF format following the author packages link from www.ams.org/tex/. The author package can be obtained free of charge by sending email to pub@ams.org (Internet) or from the Publication Division, American Mathematical Society, 201 Charles St., Providence, RI 02904, USA. When requesting an author package, please specify $\mathcal{AMS}$-LaTeX or $\mathcal{AMS}$-TeX, Macintosh or IBM (3.5) format, and the publication in which your paper will appear. Please be sure to include your complete mailing address.

**Sending electronic files.** After acceptance, the source file(s) should be sent to the Providence office (this includes any TeX source file, any graphics files, and the DVI or PostScript file).

Before sending the source file, be sure you have proofread your paper carefully. The files you send must be the EXACT files used to generate the proof copy that was accepted for publication. For all publications, authors are required to send a printed copy of their paper, which exactly matches the copy approved for publication, along with any graphics that will appear in the paper.

TeX files may be submitted by email, FTP, or on diskette. The DVI file(s) and PostScript files should be submitted only by FTP or on diskette unless they are encoded properly to submit through email. (DVI files are binary and PostScript files tend to be very large.)

Electronically prepared manuscripts can be sent via email to pub-submit@ams.org (Internet). The subject line of the message should include the publication code to identify it as a Memoir. TeX source files, DVI files, and PostScript files can be transferred over the Internet by FTP to the Internet node e-math.ams.org (130.44.1.100).

**Electronic graphics.** Comprehensive instructions on preparing graphics are available at www.ams.org/jourhtml/graphics.html. A few of the major requirements are given here.

Submit files for graphics as EPS (Encapsulated PostScript) files. This includes graphics originated via a graphics application as well as scanned photographs or other computer-generated images. If this is not possible, TIFF files are acceptable as long as they can be opened in Adobe Photoshop or Illustrator. No matter what method was used to produce the graphic, it is necessary to provide a paper copy to the AMS.

Authors using graphics packages for the creation of electronic art should also avoid the use of any lines thinner than 0.5 points in width. Many graphics packages allow the user to specify a "hairline" for a very thin line. Hairlines often look acceptable when proofed on a typical laser printer. However, when produced on a high-resolution laser imagesetter, hairlines become nearly invisible and will be lost entirely in the final printing process.

Screens should be set to values between 15% and 85%. Screens which fall outside of this range are too light or too dark to print correctly. Variations of screens within a graphic should be no less than 10%.

**Inquiries.** Any inquiries concerning a paper that has been accepted for publication should be sent directly to the Electronic Prepress Department, American Mathematical Society, 201 Charles St., Providence, RI 02904, USA.

# Editors

This journal is designed particularly for long research papers, normally at least 80 pages in length, and groups of cognate papers in pure and applied mathematics. Papers intended for publication in the *Memoirs* should be addressed to one of the following editors. In principle the Memoirs welcomes electronic submissions, and some of the editors, those whose names appear below with an asterisk (*), have indicated that they prefer them. However, editors reserve the right to request hard copies after papers have been submitted electronically. Authors are advised to make preliminary email inquiries to editors about whether they are likely to be able to handle submissions in a particular electronic form.

**Algebra** to KAREN E. SMITH, Department of Mathematics, University of Michigan, 525 University, Suite 2832, Ann Arbor, MI 48109-1109; email: `kesmith@lsa.umich.edu`

**Algebraic geometry and commutative algebra** to LAWRENCE EIN, Department of Mathematics, University of Illinois, 851 S. Morgan (M/C 249), Chicago, IL 60607-7045; email: `ein@uic.edu`

**Algebraic topology and cohomology of groups** to STEWART PRIDDY, Department of Mathematics, Northwestern University, 2033 Sheridan Road, Evanston, IL 60208-2730; email: `priddy@math.nwu.edu`

**Combinatorics and Lie theory** to SERGEY FOMIN, Department of Mathematics, University of Michigan, Ann Arbor, Michigan 48109-1109; email: `fomin@umich.edu`

**Complex analysis and complex geometry** to DUONG H. PHONG, Department of Mathematics, Columbia University, 2990 Broadway, New York, NY 10027-0029; email: `phong@math.columbia.edu`

*__Differential geometry and global analysis__ to LISA C. JEFFREY, Department of Mathematics, University of Toronto, 100 St. George St., Toronto, ON Canada M5S 3G3; email: `jeffrey@math.toronto.edu`

**Dynamical systems and ergodic theory** to ROBERT F. WILLIAMS, Department of Mathematics, University of Texas, Austin, Texas 78712-1082; email: `bob@math.utexas.edu`

**Functional analysis and operator algebras** to DAN VOICULESCU, Department of Mathematics, University of California, Berkeley, 970 Evans Hall, Floor 9, Berkeley, CA 94720-0001; email: `dvv@math.berkeley.edu`

**Geometric topology, knot theory and hyperbolic geometry** to ABIGAIL A. THOMPSON, Department of Mathematics, University of California, Davis, Davis, CA 95616-5224; email: `thompson@math.ucdavis.edu`

**Harmonic analysis** to ALEXANDER NAGEL, Department of Mathematics, University of Wisconsin, 480 Lincoln Drive, Madison, WI 53706-1313; email: `nagel@math.wisc.edu`

**Harmonic analysis, representation theory, and Lie theory** to ROBERT J. STANTON, Department of Mathematics, The Ohio State University, 231 West 18th Avenue, Columbus, OH 43210-1174; email: `stanton@math.ohio-state.edu`

*__Logic__ to THEODORE SLAMAN, Department of Mathematics, University of California, Berkeley, CA 94720-3840; email: `slaman@math.berkeley.edu`

**Number theory** to HAROLD G. DIAMOND, Department of Mathematics, University of Illinois, 1409 W. Green St., Urbana, IL 61801-2917; email: `diamond@math.uiuc.edu`

*__Ordinary differential equations, and applied mathematics__ to PETER W. BATES, Department of Mathematics, Michigan State University, East Lansing, MI 48824-1027; email: `peter@math.msu.edu`

*__Partial differential equations__ to PATRICIA E. BAUMAN, Department of Mathematics, Purdue University, West Lafayette, IN 47907-1395' email: `bauman@math.purdue.edu`

*__Probability and statistics__ to KRZYSZTOF BURDZY, Department of Mathematics, University of Washington, Box 354350, Seattle, Washington 98195-4350; email: `burdzy@math.washington.edu`

*__Real analysis and partial differential equations__ to DANIEL TATARU, Department of Mathematics, University of California, Berkeley, Berkeley, CA 94720; email: `tataru@math.berkeley.edu`

**All other communications to the editors** should be addressed to the Managing Editor, WILLIAM BECKNER, Department of Mathematics, University of Texas, Austin, TX 78712-1082; email: `beckner@math.utexas.edu`.

# Titles in This Series

783 **Ethan Akin, Mike Hurley, and Judy A. Kennedy,** Dynamics of topologically generic homeomorphisms, 2003

782 **Masaaki Furusawa and Joseph A. Shalika,** On central critical values of the degree four $L$-functions for GSp(4): The Fundamental Lemma, 2003

781 **Marcin Bownik,** Anisotropic Hardy spaces and wavelets, 2003

780 **S. Marmi and D. Sauzin,** Quasianalytic monogenic solutions of a cohomological equation, 2003

779 **Hansjörg Geiges,** $h$-principles and flexibility in geometry, 2003

778 **David B. Massey,** Numerical control over complex analytic singularities, 2003

777 **Robert Lauter,** Pseudodifferential analysis on conformally compact spaces, 2003

776 **U. Haagerup, H. P. Rosenthal, and F. A. Sukochev,** Banach embedding properties of non-commutative $L^p$-spaces, 2003

775 **P. Lochak, J.-P. Marco, and D. Sauzin,** On the splitting of invariant manifolds in multidimensional near-integrable Hamiltonian systems, 2003

774 **Kai A. Behrend,** Derived $\ell$-adic categories for algebraic stacks, 2003

773 **Robert M. Guralnick, Peter Müller, and Jan Saxl,** The rational function analogue of a question of Schur and exceptionality of permutation representations, 2003

772 **Katrina Barron,** The moduli space of $N = 1$ superspheres with tubes and the sewing operation, 2003

771 **Shigenori Matsumoto,** Affine flows on 3-manifolds, 2003

770 **W. N. Everitt and L. Markus,** Elliptic partial differential operators and symplectic algebra, 2003

769 **Jie Wu,** Homotopy theory of the suspensions of the projective plane, 2003

768 **R. Höpfner and E. Löcherbach,** Limit theorems for null recurrent Markov processes, 2003

767 **Po Hu,** $S$-modules in the category of schemes, 2003

766 **Su Gao and Alexander S. Kechris,** On the classification of Polish metric spaces up to isometry, 2003

765 **Robert Bieri and Ross Geoghegan,** Connectivity properties of group actions on non-positively curved spaces, 2003

764 **J. Spandaw,** Noether-Lefschetz problems for degeneracy loci, 2003

763 **Yasuyuki Kachi and Eiichi Sato,** Segre's reflexivity and an inductive characterization os hyperquadrics, 2002

762 **Leiba Rodman, Ilya M. Spitkovsky, and Hugo Woerdeman,** Abstract band method via factorization, positive and band extensions of multivariable almost periodic matrix functions, and spectral estimation, 2002

761 **Oliver Druet and Emmanuel Hebey,** The $AB$ program in geometric analysis : Sharp Sobolev inequalities and related problems, 2002

760 **Markus Banagl,** Extending intersection homology type invarients to non-Witt spaces, 2002

759 **Donald M. Davis,** From representation theory to homotopy groups, 2002

758 **Alan Forrest, John Hunton, and Johannes Kellendonk,** Topological invariants for projection method patterns, 2002

757 **Douglas Bowman,** $q$-difference operators, orthogonal polynomials, and symmetric expansions, 2002

756 **José Ignacio Cogolludo-Agustín,** Topological invariants of the complement to arrangements of rational plane curves, 2002

755 **M. A. Mandell and J. P. May,** Equivariant orthogonal spectra and $S$-modules, 2002

## TITLES IN THIS SERIES

- 754 **Edward L. Green, Idun Reiten, and Øyvind Solberg,** Dualities on generalized Koszul algebras, 2002
- 753 **Daniel Panazzolo,** Desingularization of nilpotent singularities in families of planar vector fields, 2002
- 752 **Linus Kramer,** Homogeneous spaces, Tits buildings, and isoparametric hypersurfaces, 2002
- 751 **Bruce Allison, Georgia Benkart, and Yun Gao,** Lie algebras graded by the root systems $BC_r$, $r \geq 2$, 2002
- 750 **Masaki Izumi and Hideki Kosaki,** Kac algebras arising from composition of subfactors: General theory and classification, 2002
- 749 **Nanhua Xi,** The based ring of two-sided cells of affine Weyl groups of type $\tilde{A}_{n-1}$, 2002
- 748 **Jürgen Ritter and Alfred Weiss,** The lifted root number conjecture and Iwasawa theory, 2002
- 747 **Armand Borel, Robert Friedman, and John W. Morgan,** Almost commuting elements in compact Lie groups, 2002
- 746 **Peter Niemann,** Some generalized Kac-Moody algebras with known root multiplicities, 2002
- 745 **Mikhail A. Lifshits and Werner Linde,** Approximation and entropy numbers of Volterra operators with application to Brownian motion, 2002
- 744 **Roger Chalkley,** Basic global relative invariants for homogeneous linear differential equations, 2002
- 743 **Heng Sun,** Spectral decomposition of a covering of $GL(r)$: the Borel case, 2002
- 742 **J. E. Gilbert, Y. S. Han, J. A. Hogan, J. D. Lakey, D. Weiland, and G. Weiss,** Smooth molecular functions and singular integral operators, 2002
- 741 **Francisco Santos,** Triangulations of oriented matroids, 2002
- 740 **Rick Durrett,** Mutual invadability implies coexistence in spatial models, 2002
- 739 **Georgios K. Alexopoulos,** Sub-Laplacians with drift on Lie groups of polynomial volume growth, 2002
- 738 **Yasuro Gon,** Generalized Whittaker functions on $SU(2,2)$ with respect to the Siegel parabolic subgroup, 2002
- 737 **Arjen Doelman, Robert A. Gardner, and Tasso J. Kaper,** A stability index analysis of 1-D patterns of the Gray-Scott model, 2002
- 736 **Wojciech Chachólski and Jérôme Scherer,** Homotopy theory of diagrams, 2002
- 735 **Martina Brück, Xi Du, Joonsang Park, and Chuu-Lian Terng,** The submanifold geometries associated to Grassmannian systems, 2002
- 734 **Michel Van den Bergh,** Blowing up of non-commutative smooth surfaces, 2001
- 733 **Milé Krajčevski,** Tilings of the plane, hyperbolic groups and small cancellation conditions, 2001
- 732 **Jan O. Kleppe, Juan C. Migliore, Rosa Miró-Roig, Uwe Nagel, and Chris Peterson,** Gorenstein liaison, complete intersection liaison invariants and unobstructedness, 2001
- 731 **Jesús Bastero, Mario Milman, and Francisco J. Ruiz,** On the connection between weighted norm inequalities, commutators and real interpolation, 2001
- 730 **Suhyoung Choi,** The decomposition and classification of radiant affine 3-manifolds, 2001
- 729 **Michael Grosser, Eva Farkas, Michael Kunzinger, and Roland Steinbauer,** On the foundations of nonlinear generalized functions I and II, 2001

For a complete list of titles in this series, visit the
AMS Bookstore at **www.ams.org/bookstore/**.